Advance praise for

Rainwater Harvesting
for Drylands
Volume 1

"Brad Lancaster is one of those rare individuals who combines a practical ability to design and implement common-sense solutions to rainwater management issues with a clear ecological and political vision of the importance of doing so. In *Rainwater Harvesting for Drylands* Brad shows us how to use rainwater around our homes and in our communities so that our human-created landscapes reflect the abundance of nature. As we move from assumptions of scarcity to participation in abundance, our lives and our communities can be transformed."

—David Confer, Ph.D., environmental engineer and sustainable design and development consultant

"What a wonderful, enthusiastic book. Brad Lancaster lives what he preaches—a water-careful lifestyle that is all about more life. Brad is a worthy teacher—his love and deep respect for water shines through on every page."

—Ben Haggard, author, sustainable systems designer, and permaculture teacher

"Brad Lancaster's *Rainwater Harvesting for Drylands,* is an important book. Its teachings should not just be applied to drylands. It's about using hydrological cycles to create and support sustainable landscapes, and the lessons are universal and useful wherever you live.
This book is where to start with environmental restoration. His story of "The Man Who Farms Water" in Africa is a microcosm and metaphor for the brilliant use of Nature's operating instructions. Most highly recommended!"

—John Todd, Ph.D., Research Professor and Distinguished Lecturer, The Rubenstein School of Environment and Natural Resources, The University of Vermont; President, Ocean Arks International

"This important and timely water-harvesting book reads like a conversation with a trusted friend. As such, it is an effective how-to and why-for manual for living within our means in our shared watersheds.
Heartfelt thanks, Brad, for spotlighting the route to abundance in these arid climes!"

—Barbara Clark, project manager, Teran Watershed Project, Cascabel AZ

See last pages of book and www.HarvestingRainwater.com for more testimonials.

green press INITIATIVE

Rainsource Press is committed to preserving ancient forests and natural resources. We elected to print *Rainwater Harvesting for Drylands, Volume 1.* on 50% post consumer recycled paper, processed chlorine free. As a result, for this printing, and the last printing, we have saved:

32 trees (40' tall and 6-8" diameter)
13,760 gallons of water
5,534 kilowatt hours of electricity
1,516 pounds of solid waste
2,979 pounds of greenhouse gases

Rainsource Press made this paper choice because our printer, Thomson-Shore, Inc., is a member of Green Press Initiative, a nonprofit program dedicated to supporting authors, publishers, and suppliers in their efforts to reduce their use of fiber obtained from endangered forests.

For more information, visit www.greenpressinitiative.org

THE REGENERATIVE FUND

Once production costs are recovered, 10% of the profits from each book sold will go to the "Regenerative Fund." This money will help fund projects, presentations, and publications that further promote rainwater harvesting and other sustainable living strategies. The fund will also be used to help distribute these resources to libraries and schools.

Rainwater Harvesting
for Drylands
Volume 1

GUIDING PRINCIPLES TO WELCOME RAIN
INTO YOUR LIFE AND LANDSCAPE

Brad Lancaster

Illustrated by
Joe Marshall
Silvia Rayces
Ann Phillips
Roxanne Swentzell
Gavin Troy

RAINSOURCE PRESS

Tucson, Arizona
www.HarvestingRainwater.com

Published by:
Rainsource Press
813 N. 9th Ave.
Tucson, AZ 85705
U.S.A.
www.HarvestingRainwater.com

First Edition
Printed and bound in the United States of America on acid-free, recycled paper

Cover design: Kay Sather
Front cover art: Gavin Troy
Back cover illustration: Joe Marshall, www.planetnameddesire.com
Back cover photo: Josephine Thomason
Book design: Teri Reindl Bingham
Illustrations: Joe Marshall, Silvia Rayces, Ann Phillips, Roxanne Swentzell
Photographs: Unless otherwise noted, all photographs are by Brad Lancaster

Lancaster, Brad
    Rainwater Harvesting for Drylands, Volume 1: Guiding Principles to
    Welcome Rain Into Your Life and Landscape/Brad Lancaster.
    p.   cm.
    Includes bibliographical references and index.
    ISBN 0-9772464-0-X
    1. Rainwater harvesting. 2. Water harvesting. 3. Landscape design.
    4. Ecology. 5. Sustainable development 6. Do-it-yourself technology

Library of Congress Control Number (LCCN): 2005907763

To

My parents Stew and Diana Lancaster,
and my grandparents Herb and Martha Lancaster and
Frits and Jean van't Hoogerhuijs for having always
encouraged me to pursue my wild ideas and dreams

Mr. Zephaniah Phiri Maseko and family
for enabling me to see the whole

All the stewards of the earth
who teach by the example they live

The Eight Principles of Successful Rainwater Harvesting

1. Begin with long and thoughtful observation.

Use all your senses to see where the water flows and how. What is working, what is not? Build on what works.

2. Start at the top (highpoint) of your watershed and work your way down.

Water travels downhill, so collect water at your high points for more immediate infiltration and easy gravity-fed distribution. Start at the top where there is less volume and velocity of water.

3. Start small and simple.

Work at the human scale so you can build and repair everything. Many small strategies are far more effective than one big one when you are trying to infiltrate water into the soil.

4. Spread and infiltrate the flow of water.

Rather than having water erosively runoff the land's surface, encourage it to stick around, "walk" around, and infiltrate into the soil. Slow it, spread it, sink it.

5. Always plan an overflow route, and manage that overflow as a resource.

Always have an overflow route for the water in times of extra heavy rains, and where possible, use that overflow as a resource.

6. Maximize living and organic groundcover.

Create a living sponge so the harvested water is used to create more resources, while the soil's ability to infiltrate and hold water steadily improves.

7. Maximize beneficial relationships and efficiency by "stacking functions."

Get your water harvesting strategies to do more than hold water. Berms can double as high and dry raised paths. Plantings can be placed to cool buildings. Vegetation can be selected to provide food.

8. Continually reassess your system: the "feedback loop."

Observe how your work affects the site—beginning again with the first principle. Make any needed changes, using the principles to guide you.

Contents

List of Illustrations

Boxed Information

Acknowledgments

This book never would have been written, let alone published, without the incredible help and motivation from many water harvesters, friends, family, neighbors, professionals, teachers, and students. I am indebted to you all, you are the community that supports and nourishes me, and part of a larger force I see striving to enhance the quality of life for everyone. Here, I list a few of the many that helped me, though my heartfelt thanks goes out to you all.

Thank you Mr. Zephaniah Phiri Maseko and family for your inspirational life and stories. For your teachings, input, encouragement, guiding examples, and ideas thanks to Tim Murphy, Vicki Marvick, Ben Haggard, and Joel Glanzberg of Regenesis Group; Barbara Rose, Rocky Brittain, Chris Meuli, Mary Zemack; Jeremiah Kidd of San Isidoro Permaculture, Nate Downey and Melissa McDonald of Santa Fe Permaculture, Richard Jennings of Earthwrights Designs, Art Ludwig of Oasis Design, Russ Buhrow, Dan and Karen Howell, Bob and Pamela Mang, Dave Tagget, Steve Kemble and Carol Escott, Bert Lopez, Jim and Karen Brooks, Meg Keoppen, Jeff Blau, Toby Hemenway, David Omick and Pearl Mast, the staff of the Botswana Permaculture Trust, the staff of the Fambidzinai Permaculture center, Justin Willie and the other staff of the Black Mesa Permaculture Project. Thank you Bill Mollison and David Holmgren for creating a "big umbrella" framework within which I can strive for and articulate integrated systems.

Much gratitude to all peer reviewers of the draft manuscripts for your time and wise editing, comments, and suggestions. To David Aguirre, Tony Novelli, and Steve Malone for keeping my computer going, and Josh Schachter for the generous use of your slide scanner and computer. Thank you to Joe Marshall, Silvia Rayces, Ann Phillips, Gavin Troy, Roxanne Swentzell, and Kay Sather for your beautiful art that brings ideas to life and vision. And to Gary Paul Nabhan for your glowing foreword and visionary work.

Then there is that core of noble, amazing souls that I used endlessly as sounding boards for the book's concept, content, and style: Eileen Alduenda and Ann Phillips—without your incredible organizational skills, editing, understanding, management, and support this book never could have existed—really you are co-authors. Anastasia Rabin, you've kept my passion alight (in many ways), endured never-ending rainbook conversations, and kept the project true to its intent. Brock Dolman, your word play, enthusiasm, and spirit of fun gave life to abundant waterspread. Kevin Moore, your friendship and support pulled me and my writing out of serious bogs. David Confer, you've added invaluable data, diverse perspectives, and steady support, and along with Wayne Moody, have provided the opportunity for many of this book's ideas to be realized.

Thanks to editors Eileen Alduenda, Ann Phillips, Shay Soloman, Frank McGee, Tom Brightman, Matt Weber, Mac Hudson, Dan and Shelly Dorsey, and most recently, skillfully, and thoroughly, June Fritchman. And thanks to many of you, especially June for helping me refine and divide the book into tastier and more easily digestible portions. Thanks to Kaye Sather for editing the illustrations, and Brandy Winters for early illustration work. Teri Bingham, thank you for your beautiful book design and layout.

Thanks to the many people helping provide key information and research such as the City of Tucson and University of Arizona library staff, Brian Barbaris from the University of Arizona's Department of Atmospheric Sciences, University of Arizona professors Tom Wilson and Jim Riley for patient help with soils and text review, Virginia Welford and Christina Bickelmann of the Arizona Department of Water Resources for water use and conservation data, and Frank Sousa for Tucson stormwater data and helping make Tucson's water-harvesting guide happen.

Thanks everyone who invested in my vision for this book and helped raise the capital for its printing through your preorders. You had to endure delay after delay. I hope your expectations and needs have been met or exceeded.

I am indebted to all who provided me with testimonials, and want you to know that it was your great work in the world that lured me to ask for your thoughts.

Thank you, Mom and Dad, for consistently encouraging me in the pursuit of my passions. Thank you, brother Rodd, for going with, and helping me build the vision, for our site.

Finally, I am indebted to the indigenous dryland cultures who knew—and know how to—sustainably live with the land. May that knowledge never be forgotten.

Foreword

By Gary Paul Nabhan

Although rainwater harvesting has been accomplished by humans in virtually every drought-vulnerable region of the world for millennia, our society seems to have some collective amnesia about the utility, efficiency, sustainability, and beauty of these time-tried practices. Fortunately, this book and Brad's lifelong passion for practical, ecological, and aesthetically pleasing solutions to our water woes may cure us of that amnesia just when we most desperately need to remember such solutions are readily at hand. From where I write this in Northern Arizona, nine out of every ten trees outside my window are dead, due to the worst drought in fourteen hundred years, and the artificial reservoir known as Lake Powell is projected to go dry within six more years. And yet, those of my neighbors who harvest water off their roofs, parking lots, or slopes (as we do) have never had to haul in water during the last six years of subnormal precipitation, and elderly Hopi farmers have still produced crops every year in the floodwater (ak-chin) fields. At a time when surface- and ground-water is becoming increasingly privatized, fought over, and transferred between watersheds and aquifers as if it were but one more globalized commodity, Brad demonstrates a diversity of strategies that can quench our thirst, sustain local food production, and keep peace among neighboring cultures. Because struggles for access to water are likely to be one of the most frequent causes for warfare and social unrest over

the next half century on every continent, Brad should be nominated for the Nobel Prize for Peace for offering the world so many elegant means of avoiding such struggles through local harvesting of both water and traditional ecological knowledge.

Like many arid land ecologists scattered around the world, I was first inspired to consider the supreme importance of water harvesting for desert cultures by reading Michael Evenari's classic, *The Negev—The Challenge of a Desert*, about Israeli Jewish attempts to learn from their ancient neighbors, the Nabateans, who drew upon diverse runoff catchments and storage practices to make their prehistoric civilization flourish at Petra, the Negev, and Sinai. With the likes of arroyo-of-consciousness journalist Chuck Bowden and straw-bale movement founder Matts Myhrman, I sought out older treatises and surviving practitioners of O'odham (Papago) ak-chin farming in the Sonoran Desert. We found that there was much to learn from our desert neighbors about the harvesting of both water and nutrients; Brad has continued and extended our earlier, haphazard efforts of rescuing such knowledge from Native American elders. But Brad has also gone two steps further than many of us. He has essentially accomplished a worldwide survey of water-harvesting practices, humbling his predecessors by compiling a dizzyingly diverse portfolio of strategies, techniques, and technologies. He has then tried and fine-tuned every one of these strategies, so that he now has firsthand knowledge of how they function, and at what cost. His own desert abode is like a walk-through encyclopedia of water-harvesting techniques gleaned from cultures and innovators from around the world.

There is both quantitatively-informed precision and beauty in what Brad has implemented, and this combination is a rarity in our modern world. Technological fixes have grown increasingly ugly, but as you can see from the drawings and photos in this masterwork, Brad's designs sing to us as they solve our water shortages.

A half century ago, Thomas Merton prophesized that "some day, they will even try to sell you the rain," warning us that the privatization and corporate control of our hydrological destiny could become our doom. What Brad's genius safeguards for us is "water democracy," and I predict that this concept will become a keystone of environmental justice throughout the desert regions of the world, if not everywhere. We will no longer think of desert living as "lacking," or "limited," but celebrate the abundance before us. With tongue in cheek, we may even offer our sympathies to those who live in soggy, "drought-deficient" places, who may never be able to share the joy with us of harvesting our own fresh, delicious water, just as horned lizards have done off their very own backs since they first emerged on this dry planet. Blessings to you Brother Brad, the Patron Saint of Water Democracy.

Gary Paul Nabhan is the desert rat author of *The Desert Smells Like Rain* and *Coming Home to Eat*, and Director of the Center for Sustainable Environments. Although he has a Ph.D. in arid land resources, he has learned more from Brother Brad than all his professors combined.

Introduction

Catch rain where rain falls.

—East Indian proverb

I love the rain! I love to drink it, sing in it, dance in it, bathe in it. Of course that's only natural; our bodies are more than 70% water. You and I and everyone else—we're walkin', talkin', *rain.*

Rain is the embodiment of life. It infuses water into our springs, rivers, and aquifers. It cools us, greens the land, and nourishes the plants that feed us. It cleans the air, washes salts from the soil, and makes the animals sing.

Yet the world's supply of fresh water is finite. Less than one half of one percent of all the water on Earth is fresh and available. The rest is seawater, or frozen. Our supply is renewed only through precipitation, a precious gift from the sky that falls as droplets, hail, or snowflakes, and then flows over the landscape as runoff. In this book, I refer to the gift as "rainwater." And the gift is ripe for harvesting.

Rainwater harvesting captures precipitation and uses it as close as possible to where it falls. The process mimics intact and healthy ecosystems, which naturally infiltrate rainwater into the soil and cycle it through myriad life forms. Instead of sealing and dehydrating the landscape with impervious pavement and convex shapes that drain the gift away, as most modern cities, suburbs, and home landscapes do, harvesting accepts rain and allows it to follow its natural path to productivity.

This book provides you with a simple series of integrated strategies for creating water-harvesting "nets" which allow rainwater to permeate and enhance our landscapes, gardens, yards, parks, farms, and ranches. Small-scale strategies are the most effective and the least expensive, so they are emphasized here. They're also the safest and easiest to accomplish. *They can empower you to become water self-sufficient.*

The benefits are many. By harvesting rainwater within the soil and vegetation—*in* the land, or in cisterns that will later irrigate the land, we can decrease erosion, reduce flooding, minimize water pollution, and prevent mosquito breeding (within water standing on top of the soil for more than three days). The process also generates an impressive array of resources: It can provide drinking water, generate high quality irrigation water, support vegetation as living air conditioners and filters, lower utility bills, enhance soil fertility, grow food and beauty, increase local water resources, reduce demand for groundwater, boost wildlife habitat, and endow us and our community with skills of self-reliance and cooperation!

MY RAINWATER-HARVESTING EVOLUTION

In 1994, my brother Rodd and I began harvesting water in our backyard by digging, then mulching a basin around a single drought-stressed sour orange tree. We graded the soil around the basin so runoff from the surrounding area, and the neighbor's roof would drain to the tree. The results amazed us. After a single rain, the tree burst out with new leaves, a dreamy show of fragrant blossoms, and an abundant crop of fruit that was soon converted into tasty marmalade and "orangeade" by family, friends, and neighbors. That was ten years ago, and we've since kept our irrigation of that tree to just three supplemental waterings per year. Yet we live within the Sonoran Desert where annual rainfall averages just 12 inches (304 mm), and most folks water their citrus trees at least once a week.

With the citrus tree flourishing, we decided to mimic its success and make rainwater the primary water source for all our outdoor needs. Using methods described in chapters 3 and 4, and more in depth in volume 2, we created and planted undulating water-harvesting earthworks throughout our once barren yard. The rain then gently soaked into the soil, soil erosion ceased, and verdant life began sprouting everywhere. We planted shade trees that grew tall around the house, lowering summer temperatures enough for us to eliminate our evaporative cooler (improved insulation, painting the house's exterior white, and passive ventilation also helped). We then boosted the growth of these trees still further using greywater recycled from the drains of our home's sinks, shower, and washing machine. Our daily municipal water use dropped from the Tucson residential average of 114 gallons (431 liters) per person per day[1] to less than 20 gallons (75 liters) per person per day, and our water and electric bills plummeted. This earned us five visits from workers at both the water and electric utilities because they were sure our meters were broken.

We wanted to do more. Every time it rained our street turned into a river, fed by runoff from neighborhood roofs, yards, and pavement. We redirected that runoff to 19 young native trees we planted in the barren public right-of-way adjacent to our property. These low-water-use trees now sing with nesting songbirds and offer a beautiful shaded canopy for pedestrians, bicyclists, and motorists. Water that once flowed away now supports trees that filter pollutants carried in the road's runoff as they shade and cool the street (see the chapter on reducing hardscape and creating permeable paving in volume 2 for more details). Mosquito populations have plunged because water no

longer stands in puddles, but is instead soaked up by spongy mulch and taken up by plants.

Our lot was once hot, barren and eroded, with a house that could only be made comfortable by paying to mechanically alter its climate. Now our yard is an oasis producing 15 to 25% of our food, and after growing trees and installing solar panels to power fans, we no longer pay a cent to heat and cool our home (keep in mind we are also the type that will put on a sweater before firing up a wood stove). We've switched from contributing to neighborhood flooding to contributing to neighborhood flood control, and our landscape enhances local water resources instead of depleting them. On our 1/8-acre (0.05-ha) lot and surrounding right-of-way we currently harvest annually over 100,000 gallons (454,600 liters) of rainwater within a 1,200 gallon tank, the soil, and vegetation, while using less than 20,000 gallons (75,600 liters) of municipal groundwater for our domestic needs and landscape irrigation in dry spells. Four-fifths of the water we now use comes from our own yard, not from city supply.

When friends and neighbors drop by they see the potential of water harvesting and learn how to do it themselves. Many then go home and spread the "seeds" by setting up work parties and creating their own rain-fed oases. That, in essence, is my vision: Harvest rainwater within our own yards and neighborhoods, encourage emulation, enhance rather than deplete our water resources, and improve the lives of everyone in our community.

This has become my passion and my profession. Fueled by what I've learned from hands-on experience, I've taught countless workshops on rainwater harvesting and permaculture—an integrated system of sustainable design. I've designed and consulted on self-sustainable water-harvesting strategies and systems for many backyard gardeners, neighborhoods, city projects, land restoration endeavors, and major housing developments. And all these projects and teachings are based on the principles I give you in this book. It is my hope to plant more water-harvesting "seeds."

If every neighborhood in my hometown, and yours, harvested rainwater in an integrated way we could greatly reduce the need for our concrete-clad, water-draining, multi-million dollar flood control infrastructure. Our yards and public right-of-ways would become a new tree-lined, water-harvesting *greenfrastructure*, no longer requiring us to spend billions extracting and importing water from other communities to supplement our drained and dwindling supply.

The economics speak for themselves. By valuing and harvesting the ignored resource of rain, groundwater levels can stabilize and even rise again, failing springs and creeks can come back to life, native plants can recolonize wasteland, and ultimately the global hydrologic cycle can benefit, while we simultaneously reduce our cost of living! Real life examples of these scenarios permeate this book, encouraging us to think globally as we act locally.

How do we get there? We become aware, apply our awareness, and throw down the welcome mat to invite rainwater into our lives and landscape.

WHO THIS BOOK IS FOR

This book, along with volumes 2 and 3, is for anyone who wants to harvest rainwater in a safe, productive, sustainable way. You can be the expert and steward of your land, whether you live on an urban or rural site, big or small. This book explains what water harvesting is, how to do it, and how to apply it to the unique conditions of your site. The aim is to realize the maximum effectiveness for the least effort and cost. You'll be guided in the design of new water-harvesting landscapes or in the retrofit of ones that exist.

This book will also help you convey water-harvesting ideas to the landscape designers and maintenance workers who may be helping you at your site. Planners and designers will discover how to devise more efficient strategies and integrated environments appropriate for dryland communities as well as those with abundant water. Landscapers and gardeners will learn how to create and maintain water-harvesting earthworks. Activists will learn how water-harvesting projects can bring people together, create a sense of place, and empower the community.

Dryland-appropriate strategies are emphasized throughout this book, because this is where the need is the greatest (see box I.1). Many are borrowed from, or based on, traditions that have allowed people to

---

Box I.1. Drylands: A Definition

Drylands are typically defined as areas of the world where potential average yearly moisture loss (evapotranspiration) exceeds average yearly moisture gain (precipitation). Evapotranspiration is the combined measurement of water loss to evaporation and transpiration.[2] Transpiration is the loss of moisture from plants to the air via the stomata within their leaves.

More than 6.1 billion hectares, 47.2% of the Earth's land surface, is dryland. A fifth of the world's population lives in dryland habitat.[3] Normal dry seasons can last six months or more. Droughts can last for years.

---

survive and thrive in arid environments for thousands of years (appendix 2). Yet the principles are universally applicable; wet and dry climates are both susceptible to drought and flooding. Rainwater harvesting reduces the impacts of dry seasons, droughts, and floods. By optimizing the capture of the rain we buffer our lands from changing climates and climatic extremes, while making our lands more resilient.

My goal is to enable you to appreciate the value of rainwater and begin to use it as your primary water source—if not for the entire household, at least for your landscape. You'll not only get the most from rainfall, no matter how scarce, but from other water sources as well. An integrated landscape harvests all water and such resources as topsoil, organic matter, and nutrients. It acts as a concave, life-giving sponge rather than a convex, eroding burial mound, which drains water and other resources away. (See figure I.1.)

HOW TO USE THIS BOOK AND VOLUMES 2 AND 3

From the start, I've intended *Rainwater Harvesting for Drylands* to be an all-in-one source on how to conceptualize, design, and implement integrated and sustainable rainwater-harvesting systems. And it has grown and grown with ongoing research, experience, insights, and exposure to the great work of others. *The result was the resource I've always wanted!* But it took the form of a massive single volume too intimidating in size to the

Fig. I.1A. A landscape draining resources.
Arrows denote runoff flow.

Fig. I.1B. A landscape harvesting resources.
Arrows denote runoff flow.

uninitiated and too large to easily carry while observing a site, brainstorming design ideas, or implementing the plan. So, I've divided the book up into three user-friendly, more portable volumes. I strongly recommend everyone read volume 1, since it puts all three volumes in context and lays down the foundation of how to conceptualize a truly efficient and productive integrated system that can do far more than just harvest rainwater. Volumes 2 and 3 then expand on this by elaborating on how to employ the specific techniques that flesh out and realize the general strategies presented in volume 1. Volume 2 focuses on earthworks passively harvesting rainwater and greywater within the landscape. Volume 3 focuses on roof catchment and cistern systems.

Here's a more detailed breakdown:

VOLUME 1

The Introduction makes the case for harvesting rainwater and shifting to a paradigm of more sustainable water management.

The following chapters in this volume then lay out the steps for creating an integrated water-harvesting system:

Chapter 1 is intended to help you conceptualize the basic water-harvesting principles that will enable you to create a system that maximizes safety, efficiency, and productivity. *This chapter is the core of the book and the heart of successful water harvesting.*

In chapter 2, you will walk your watershed and assess your site's water resources.

Chapters 3 and 4 are intended to be an overview of the kinds of techniques you can use.

Chapter 3 is a discussion which will determine which of the water-harvesting strategies (earthworks, cisterns, or both) would be best for your site and needs. It also provides an overview of and illustrations of various earthworks techniques, and some illustrations and discussion of tanks.

Chapter 4 will discuss integrating other on-site resources into your system. Here you can plan how to get more than rainwater for your harvesting efforts, and I describe how my brother and I have done so on our site.

There are several appendices. Appendix 1 shows erosion patterns and earthwork techniques that may be used to remedy them. Appendix 2, by Joel Glanzberg, is about traditional Native American water-harvesting techniques in the Southwest. Appendix 3 gives you many water-harvesting calculations. Appendix 4 provides a list of example plants and their water requirements, and while this list is specifically for Tucson, Arizona, other readers may find it useful. Appendix 5 provides sample worksheets for figuring on-site water resources, water budgets, etc., and is intended as a structure to write down your observations and calculations for future reference. Appendix 6 compiles the resources referenced throughout the book's text, as well as providing many other useful resources: books, films, and websites. There are additional resources about organizations promoting rainwater harvesting and permaculture at a community level.

There are also reference notes, a glossary, and an index.

VOLUME 2: EARTHWORKS

In this volume, you will learn how to select, place, and construct your chosen water-harvesting earthworks. It presents detailed how-to information and variations of all the earthworks, including chapters on mulch, vegetation, and greywater recycling so you can customize the techniques to the unique requirements of your site.

VOLUME 3: ROOF CATCHMENT AND CISTERN SYSTEMS

Here, you will learn to select, size, design, build or buy, and install your chosen roof catchment and cistern systems. Principles unique to cistern systems are presented along with numerous tank options, and design strategies that enable your tank to do more than harvest water.

Real life stories of people creating and living with water-harvesting landscapes and systems frame all three volumes. We have honed our skills through countless hours of hands-on design, implementation, maintenance, and living with our systems. The scale and context of some of the systems presented may seem too large, too small, too urban, or too rural to apply to your site, but keep in mind that if you grasp how the principles and ethics have been realized in the various systems, you can adapt them to the scale and context your site requires.

THE VALUE OF RAINWATER

*Don't pray for rain, if you can't
take care of what you get.*

—R. E. Dixon (1937) Superintendent, Texas
Agricultural Experiment Station, Spur, Texas

So, you want to harvest rainwater—right on! Let's celebrate the value of rainwater and the many water resources it supports, because how we value our water resources directly relates to how we perceive, utilize, and *manage* them.

Precipitation (rain, hail, sleet, and snowfall) is the primary source of fresh water within our planet's hydrologic cycle. This precipitation, or "rain," supplies all secondary sources of water, including groundwater and surface water in creeks, rivers, and lakes. If consistently pumped or drained faster than they are replenished, these secondary sources eventually cease to exist.

Precipitation is naturally distilled through evaporation prior to cloud formation (fig. I.2), and thus is one of our purest sources of water.[4] Rainwater has about 100 times less total dissolved solids (TDS) than ground and surface water in my hometown![5]

Rain is considered *soft* due to the lack of calcium carbonate or magnesium in solution, and is excellent for cooking, washing, and saving energy. Much of our ground and surface water is *hard* due to the calcium and magnesium compounds that dissolve as water runs through or over soil. These compounds deposit on or in cookware, pipes, and water heaters forming white "scale" that inhibits heat conduction and shortens pipe and appliance life. Using rainwater instead saves energy and maintenance costs, and can prolong the life of water heaters and pipes.[6] Rainwater use also reduces detergent and soap requirements, and eliminates soap scum, hardness deposits, and the need for a water softener (sometimes required with well water systems),[7] besides being a natural hair conditioner.

Fig. I.2. Pure rainwater

Rainwater is a natural fertilizer. According to cooperative extension agent John Begeman, rain contains sulfur—important in the formation of plant amino acids, and it contains beneficial microorganisms and mineral nutrients collected from dust in the air—important for plant growth. Rainwater also contains nitrogen, which triggers the greening of plants. During storms, lightning strikes enable atmospheric nitrogen to combine with hydrogen or oxygen to form ammonium and nitrate, two forms of nitrogen that go into solution in atmospheric moisture and can be used by plants.[8]

Rainwater has the lowest salt content of natural fresh water sources so it is a superior water source for plants. Calcium, magnesium, potassium, and sodium salts are abundant in the earth's crust. Soils with high salt concentrations inhibit plant growth by reducing vegetation's ability to take up water and conduct photosynthesis.[9] Soils high in sodium have a tendency to disperse—or lose their structure—resulting in poor water infiltration, and soil crusting, which restricts root penetration and impedes seedling emergence.[10] As David Cleveland and Daniela Soleri write in *Food From Dryland Gardens*, "Salty soils occur naturally in arid areas where not enough rain falls to wash soluble salts down and out of the root zone. Irrigation [with surface or groundwater] makes the situation worse, since surface water and groundwater contain more salt than rainwater. Salt tends to build up in the soil as water is continually added through irrigation."[11] As long as the soil drains and enough rainwater is applied, rainwater can dilute these salts and flush them out of the root zone.[12]

Rainwater comes to us free of charge. It falls from the sky and we don't pay to pump it nor do we pay a utility company to deliver it (fig. I.3).

Yet, current management of household and community water resources does not reflect the true value of rain. Rather than treating it as our primary renewable source of fresh water we typically treat rainwater as a nuisance, diverting it to the storm drain, drainage ditch, or pollutant-laden street. In its place we invest vast resources acquiring lower-quality, secondary sources of ground and surface water. Such contemporary water management contrasts sharply with rainwater-harvesting traditions.

RAINWATER HARVESTING THROUGH LAND AND TIME

Around the globe, traditions and historic evidence of rainwater harvesting illustrate its importance as a primary water source. According to John Gould and Erik Nissen-Petersen, authors of *Rainwater Catchment Systems for Domestic Supply*, the origins of rainwater collection may extend as far back in human history as the use of fire as evidenced by the traditional practices of the hunter-gatherer Kalahari Bushmen (the San Peoples) collecting, storing, and burying rainwater in ostrich eggs to be recovered months or years later.

Roof runoff was the main source of water for many Phoenician and Carthaginian settlements from the sixth century B.C. into Roman times, when harvested rainwater became the primary water source for whole cities. As far back as 2,000 years, rain-fed cisterns provided domestic water throughout North Africa, the Mediterranean, the Middle East, and Thailand. There is a 4,000-year-old tradition of rainwater-collection systems for domestic supply and agriculture throughout

Fig. I.3. Rain is always free.

the Indian subcontinent, and water harvesting in China may have extended back 6,000 years. Rooftop collection and storage of rainwater was the principal source of water in Venice, Italy from 300 to 1600 A.D. Aztec ground catchment systems were in use by 300 A.D. Native Americans in the Southwest desert used a variety of techniques (see appendix 2.) Island cultures still rely on rainwater in parts of Japan, the Caribbean, and Polynesia. The tradition of harvesting rainwater in cisterns at isolated homesteads and farms continues today in the U.S., Canada, Australia, and New Zealand.[13]

THE SHIFT AWAY FROM RAINWATER

We've moved away from these traditions over the past 150 years as new technologies have enabled us to access, pump, and transport huge volumes of groundwater and surface water: secondary water sources in the hydrologic cycle. These secondary supplies seemed infinite so we kept taking more. In 1930 there were 170 irrigation wells tapping the Ogallala aquifer that stretches 1,300 kilometers from the Texas panhandle to South Dakota; by 1959 there were over 42,000.[14] As Charles Bowden writes, "By the sixties the High Plains had 5,500,000 acres under irrigation and men were working through the night to direct the flow from the ceaseless pumps."[15]

Surface water and groundwater—secondary water sources in the hydrologic cycle—appeared to be more

Fig. I.4A. A landscape on the wasteful path to scarcity. Rain, runoff, and topsoil are quickly drained off the landscape to the street where the sediment-laden water contributes to downstream flooding and contamination. The landscape is dependent upon municipal/well water irrigation and imported fertilizer.

Fig. I.4B. A landscape on the stewardship path to abundance. Rain, runoff, leaf drop, and topsoil are harvested and utilized within the landscape contributing to flood control and enhanced water quality. The system is self-irrigating with rain and self-fertilizing with harvested organic matter.

convenient, profitable, and dependable than rain—the primary source. Surface water and groundwater became the "primary" water resources in our modern water management system. Waste became more common than conservation. We came to see rain as a source of flooding that needed to be drained away. This appeared to work for a while, but the reality of this hydro-illiteracy has hit.

SCARCITY OR ABUNDANCE

While the hydrologic cycle continuously recycles earth's water to produce renewed fresh rain, the rate at which fresh water is produced does not meet our ever-growing demand. In the face of this demand, our planet's fresh water resources are finite. Current consumption rates are lowering groundwater levels and depleting surface water flows the world over. According to the Blue Gold Report, global water consumption is doubling every 20 years—more than twice the rate of human population growth. If current trends persist, by 2025 the demand for fresh water will be 56% more than is currently available.[16] The Ogallala aquifer is being depleted eight times faster than nature can replenish it.[17]

We have reached a turning point in our water use and management. As my friend Brock Dolman says, "We can choose to be 'scared in the city' because of water scarcity, or we can choose water *abundance*—the fine and thriving condition in which 'our buns can dance'!"

While I focus primarily on rainwater, the way we value ALL water resources shapes our future. In box I.2, I lay out the tenets of two contrasting paths of water use (see also figure I.4). Read on and ask yourself: What path am I on now? What path do I want to take from now on?

DRAINING VERSUS INFILTRATING WATER

Living the wasteful path to scarcity

We **drain** our communities by diverting our rainwater *away from* rather than infiltrating it *into* our landscapes, waterways, and aquifers. We replace living nets of pervious vegetation and topsoil with *im*pervious asphalt, concrete, and buildings, inducing rainwater to rush across the land and drain out of the system.

We create landscapes of burial-like mounds (convex shapes), which drain rather than retain water, topsoil,

Box I.2. The Scarcity Path versus the Abundance Path

THE WASTEFUL PATH TO SCARCITY

- Water scarcity is the condition in which our local water supply cannot continue to meet demand because our "fresh water bank account" is being drained.

VALUES

- We don't value and appreciate water.
- We treat rainwater as a problem—a substance we must get rid of.
- We think of groundwater and surface water as infinitely available—substances we can afford to mismanage and waste.
- We think that we as humans are separate and independent from nature.

CHARACTERISTICS

- As individuals and communities, we do not take responsibility for managing our own water accounts.
- We continually draw on groundwater savings.
- We don't make new water deposits.

RESULTS

- This extractive relationship with our natural resources leads to their degradation and depletion.

THE STEWARDSHIP PATH TO ABUNDANCE

- Water abundance is the condition in which we adjust our patterns of water management and use until our locally available water supply meets and ultimately exceeds our needs.

VALUES

- We value all water, recognizing it as the basis of our living biological system.
- We treat rainwater as the foundation of the life-sustaining hydrologic cycle.
- We treat groundwater and surface water as reservoirs that naturally accumulate and concentrate rainwater, and do not waste these.
- We understand and celebrate that we as humans are part of the earth's natural system, which sustains us all.

CHARACTERISTICS

- We as individuals and as communities thoughtfully manage our own water accounts.
- We withdraw our groundwater savings only in times of true need.
- We continually make water deposits.

RESULTS

- The abundance path contributes to the regeneration of water and other renewable natural resources. We work to enhance the environment by providing the natural "compound interest" of healthy soil, plant, and animal communities that are the source of water, food, shelter, air, and beauty.

---

and organic matter. We place plants on top of these mounds, and pump water to them through an irrigation system, while rainwater drains away from the vegetation. Care packages of purchased fertilizer are applied to replace the lost topsoil and fertility. It is a system reminiscent of a hospitalized patient on an intravenous drip.

We direct roof runoff to streets and storm drains via gutters, downspouts, and landscaped river cobble "stream beds" that eject water quickly from yards. (See figure I.5A.)

Box I.3. Draining Facts

- A recent report prepared by American Rivers states that the rapid expansion of paved-over and developed land in communities all across the U.S. is making the effects of drought worse. Development in Atlanta, Georgia and surrounding counties contributes to a yearly loss of rainwater infiltration ranging from 57 to 133 billion gallons. If managed on site, this rainwater—which could support annual household needs of 1.5 to 3.6 million people—would filter through the soil to recharge aquifers, and increase underground flows to replenish rivers, streams, and lakes.[18,19]

- Twenty-five percent of the land within incorporated Tucson is covered with impervious cover such as asphalt, concrete, or buildings.[20] In higher density cities such as Los Angeles, California, over 60% of the land surface is covered with pavement.[21]

Fig. I.5A. A landscape on life support draining its resources away. Note the mounded planted areas.

Fig. I.5B. A sustainable landscape harvesting and recycling on-site resources. Note the sunken, mulched planted areas and native vegetation.

Living the stewardship path to abundance

We **infiltrate** rainwater into our soils and vegetation as close as possible to where it falls. We replace impervious surfaces with water-harvesting earthworks and tanks, and with spongy water-retaining mulch and vegetation to intercept and use runoff from sealed surfaces.

We construct bowl-like landscapes (concave shapes) to passively harvest rainwater, build topsoil, accumulate mulch, and reduce or eliminate the need for irrigation and fertilizer. This deposits the primary water source—rain—within our local soils, and reduces the need to use secondary surface water and groundwater resources.

We harvest stormwater runoff from streets into our landscapes. Streets then become passive irrigators of beautiful shade trees lining the streets and walkways. This inexpensive *greenfrastructure* reduces the

need for conventional, costly, concrete-clad storm drains. (See figure I.5B.)

OVER-EXTRACTION VERSUS CONSERVATION OF WATER

Living the Wasteful Path to Scarcity

We **over-extract** our local water sources by pumping wells and diverting water from rivers and springs faster than rainfall can naturally replenish them.

We reduce natural groundwater recharge—particularly in areas with shallow groundwater tables—by paving over surfaces and causing rapid rainfall runoff.[22]

As we drain more of our rainwater "deposits" away, we simultaneously pump and consume more of our ancient groundwater savings account. As a result, rivers dry up, water tables drop, pumping costs increase, riparian trees die, and water quality declines. (See figure I.6A.)

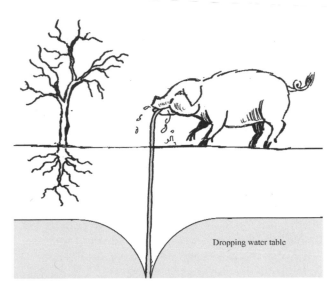

Fig. I.6A. Over-extracting groundwater

Fig. I.6B. Harvested rainwater conserving municipal/well water

Living the stewardship path to abundance

We **conserve** our fresh water resources by utilizing rainwater, recycling all water, and by mulching, using low-flow appliances, installing greywater systems, and practicing integrated design to reduce fresh water needs.

We allow natural groundwater recharge to occur by maintaining soil- and vegetation-covered landscapes and healthy watersheds and waterways.

We make deposits to our water account by allowing water to recharge and accumulate in aquifers. We use ancient groundwater supplies only during times of drought and only to meet compelling needs.

Rainwater becomes our primary water source and our groundwater "savings accounts" are enriched and reserved for times of need.

We strive to live in sustainable balance with our local water resources by living within the constraints of our site's rainwater resources/budget. (See figure I.6B.)

POLLUTING VERSUS CLEANING WATER

Living the wasteful path to scarcity

We **pollute** our finite fresh water sources.

We are disconnected from our source of water and our effects on its quality. We see water flowing from taps, delivered by massive central distribution systems, but we aren't confronted with the path water took to get to our tap.

Rain falls through the atmosphere, runs across the land surface, and infiltrates through our soils, so contamination of our air, land surface, and soil contaminates our water. We pollute our environment with

Box I.5. Over-Extraction Facts

In Arizona, groundwater pumping has dried up or degraded 90% of once-perennial streams, rivers, and riparian habitats.[24]

In California's Central Valley, overuse of ground water has collapsed the aquifer soil structure, causing a permanent loss of natural water storage capacity amounting to more than 40% of the combined storage capacity of all human-made dams and surface reservoirs in the state.[25]

In 1972, water diversion from China's once mighty Yellow River prevented it from reaching the sea for the first time in history. That year the flow did not reach the ocean on 15 days. In 1997, it failed to reach the sea on 226 days.[26]

Box I.6. Conservation Facts and Tips

The average American household could reduce water demands by 33 to 50% each year.[27] Additional water-conserving technologies and methods could cut agricultural water demands by close to 50%, and industrial water demands by 50 to 90% without sacrificing economic output or quality of life.[28]

Conserving water saves energy, and conserving energy saves water. "Power plants that use coal, oil, natural gas, or uranium also need water for steam and cooling, and they take 131 billion gallons a day—more than five times as much water as people use in their homes."[29]

Every kilowatt-hour (kWh) of coal-generated electricity uses about 1 pound of coal, produces 2 pounds of CO_2, and uses just under 0.5 gallons of water.[30,31]

Plant within your water budget. Create a landscape or garden that thrives within the constraints of your on-site rainwater and greywater resources. See chapter 2 for more.

Emphasize native plantings within your landscape. These plants have naturally adapted to survive on local rainfall patterns. They will thrive planted within water-harvesting earthworks or where given supplemental rainwater from a cistern! See volume 2, the chapter on vegetation.

sewage, pesticides, herbicides, burning of fossil fuels, chemicals dumped down drains, dripping automobile fluids, and countless other sources. (See figure I.7A.)

Living the stewardship path to abundance

We **clean** our polluted water to make it usable again.

We are in a day-to-day contact with the source of our fresh water because we see it being harvested, we maintain the surfaces it flows over, and we actively work to preserve good water quality.

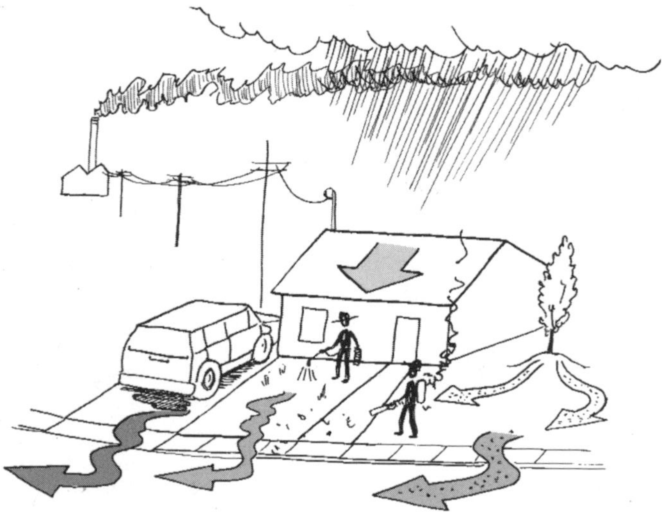

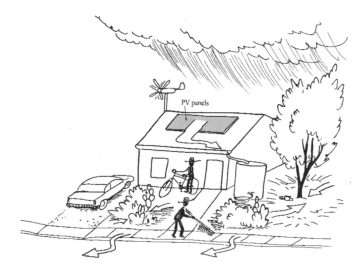

Fig. I.7A. Polluting our water and watershed at home

Fig. I.7B. Cleaning our water and watershed at home

Box I.7. Additional Water Conservation Strategies in the Home

See the Water Saver House website www.h2ouse.org for water conservation strategies around the home; see also volume 3, the chapter on determining your tank size.

WASHING MACHINES

Replacing a standard washing machine that uses 30 to 50 gallons (114–190 liters) per load of wash with a new *Energy Star*™ certified washer that uses 10 gallons (38 liters) per load of wash can reduce water consumption by 30–60% and reduce energy consumption by 50% per load.[32,33] For an average American household that's a savings of nearly 7,000 gallons (26,600 liters) a year.[34] Install a greywater system for your washer to recycle the wash water within your landscape.

PLUMBING LEAKS

Ten percent of a home's water consumption can be due to leaks. Older irrigation systems are prone to leaks, wasting over 50% to 75% of the water consumed.[35] So, regularly inspect for leaks and repair them promptly. Many municipal water companies have a free program to test for leaks.

EVAPORATIVE COOLERS, AIR CONDITIONERS, AND CONSUMPTION OF ENERGY

An evaporative cooler in Phoenix, Arizona, consumes an average 65 gallons (247 liters) per day.[36] Air conditioners don't use water on-site, but the water used to power these and other electrical appliances can be substantial if the power comes from a thermoelectric power plant (nuclear, coal, oil, natural gas, or geothermal). Passive cooling strategies found in chapter 4 can greatly reduce the need for such mechanical cooling and associated energy consumption.

OUTDOOR MISTING SYSTEMS

Outdoor misting systems use as much as 2,160 gallons (8,208 liters) per month to cool 1,000 square feet (90 m²) of patio. That's equal to over three times the average summer water use for a homeowner in Tucson, Arizona, yet a study has found that misters reduce temperatures by only 7 degrees.[37] Low-water-use native shade trees use less water than misting systems while cooling temperatures up to 20°F.[38,39]

POOLS

Multiply the surface area of a swimming pool by the local evaporation rate to determine how much water will evaporate each year. In Tucson a 400-square-foot (36-m²) pool will lose 16,000 gallons (60,800 liters) of water per year to evaporation, almost the full volume of the pool.[40] Pool covers can reduce pool water use by nearly 30%.[41] If you don't use such covers in the swimming season—at least use them in the off season. Using a community pool rather than constructing and maintaining your own can save all the water, time, money, and chemicals it takes to keep a home pool functional.

HUMAN HABITS

Consciously reduce your personal water use to reduce household water demand. For example, do not run water while brushing teeth or scrubbing hands, do not spray down driveways, patios, or yards with water, and do not run water over frozen foods to hasten thawing. While setting the example by conserving at home, push for additional conservation measures in the commercial, industrial, government, and agricultural sectors.

Box I.8. Pollution Facts

Runoff from numerous widely dispersed sources or *nonpoint-source pollution* accounts for about 60% of all surface-water pollution in the United States.[42] Nonpoint-source pollution consists of pet feces, automobile emissions, sediments and nitrogen from yards, farms and rangelands, and residual compounds from the general use of paints, plastics, etc.

One billion pounds of weed and bug poison are applied throughout the United States every year, reports *National Geographic*,[43] much of which ends up in the country's natural water systems.

Nearly 40% of U.S. rivers and streams are too polluted for fishing, swimming, or drinking.[44]

Potable water is what fills and flushes nearly all American toilets, with 6.8 billion gallons being flushed away every day.[45]

Ninety percent of the "developing" world's wastewater is still discharged untreated into local rivers and streams.[46]

We understand that the best way to have clean water is to *not pollute water in the first place*.

If we must pollute, we keep contamination to an absolute minimum and we reuse and clean that water right where we pollute it. For example, using biocompatible soaps appropriate for local soils (see volume 2 and its discussion of greywater) enables us to reuse and clean wash water or greywater on-site using the soils and plants within our landscapes.

We invest the community resources needed to purify our polluted water and return it clean to the hydrologic cycle where it can help support all life. (See figure I.7B.)

HOARDING VERSUS CYCLING WATER

Living the wasteful path to scarcity

We **hoard** water as a community by draining, over-extracting, and polluting our *local* water sources, and instead of changing our policies, lifestyles, and habits to reduce our needs, we divert water from other locations and people.

On an individual level, we watch our local water supply decline even as we buy bottles of imported spring water. We support massive dam and canal projects that divert water from others to support our demands.

Diversions and competition for water are contributing to its commodification, as water shifts from a social resource belonging to all life, to an economic commodity bought, sold, and managed by corporations that profit from the increasing scarcity of water. Water is sold to those thirsty people who have similarly squandered their water, or from whom we have taken it. (See figure I.8A.)

Fig. I.8A. Water hoarding. Large dams often hoard water from those downstream, and even other watersheds, when canal or piping systems divert water to the dams from other regions.

Living the path to abundance

We **cycle** our water as a community by increasing the productivity and potential of our limited fresh water by cycling it—using it again and again. The more life forms, uses, and resources through which fresh water *cleanly* cycles, the more life forms, uses, and resources the water can generate and support.

We stop draining, over-extracting, and polluting our local water sources by changing our policies, lifestyles, and habits to reduce water needs.

We take personal responsibility for how we treat and consume water, recognizing the negative impacts of buying imported bottled water and supporting massive water diversion schemes, and choosing instead to avoid these hoarding behaviors.

We work to prevent and reverse commodification of water by assisting people locally, nationally, and internationally to harvest rainwater, make use of grey-water, and reduce demand for secondary water sources. Water scarcity attracts market forces; water abundance does not. Water is part of the Commons (box I.12), to which we are all entitled. (See figure I.8B.)

For more information on how to protect the right to clean water for all citizens of the earth (including wildlife) see the resources in appendix 6, section B.

Fig. I.8B. Water cycling

MY COMMUNITY'S PATH

Most of Tucson's 12 inches (304 mm) of average annual rainfall pours off roofs, yards, parks, and parking lots creating torrents of street runoff that flow to storm drains (fig. I.9). We rely primarily on ancient groundwater pumped from natural aquifers beneath the city and surrounding valleys to serve municipal, agricultural, and industrial uses. In the past 100 years, overpumping has lowered this groundwater table by more than 200 feet (60 m) in some areas, and it

Fig. I.9. First Street becomes "Runoff River Street" in a summer storm.

Box I.11. Cycling Tips

We can consume just 100 gallons (454 liters) cycled through five uses, rather than consuming 500 gallons (2,273 liters) of water in a week for five different uses. And, we can harvest that 100 gallons of water from roof runoff. For example:

- **Bath water.** Harvest rainwater from roofs into a tank and bathe with it in an indoor or outdoor shower. (See chapter 4 for an example, with much more in volume 3.)

- **Irrigation.** Direct that shower greywater to shade trees and 100 gallons of shower water becomes 100 gallons of irrigation water. (See chapter 4, and volume 2, the chapter on greywater.)

- **Cooling.** Place those shade trees on the east and west sides of a building to cool air both outside and inside. This can reduce mechanical cooling that would otherwise consume 100 gallons of water in an evaporative cooler, or 100 gallons in electricity generation for an air conditioner.[55,56,57] (See chapter 4 in this volume for much more on using trees in integrated design.)

- **Food.** Select food-bearing shade trees and each 100 gallons of harvested rainwater offsets the need to use 100 gallons of water to grow food in a distant orchard. (See the plant list in appendix 4 as well as volume 2, the vegetation chapter.)

- **Fertilizer.** Collect fruit and leaves that drop to the ground around the base of the trees to create rich, water-conserving mulch, reducing the need to pump 100 additional gallons per week for irrigation. (See volume 2, the chapter on mulching.)

Box I.12. Water as Commons

I use the term "commons" as defined by Vandana Shiva in her book *Water Wars*, in which she writes, "Water is a commons because it is the ecological basis of all life and because its sustainability and equitable allocation depend on cooperation among community members."[58]

Box I.13. Water as a Human Right

Water is a limited natural resource and a public good fundamental for life and health. The human right to water is indispensable for leading a life in human dignity. It is a prerequisite for the realization of other human rights.

—United Nations, 2002[59]

continues to drop an additional 3 to 4 feet (0.9–1.2 m) more each year.[60,61] Once-perennial reaches of the Santa Cruz River and numerous springs have dried up.[62] Water quality has declined, and pumping costs have increased, as groundwater levels drop lower. Land is subsiding due to the excessive groundwater pumping.[63] Cottonwood, willow, and mesquite "bosques" or forests that used to line our waterways have died.[64] (See figure I.10, with its "before" and "after" photos.)

Pollutants from local landfills, businesses, and industry have migrated down to our aquifer, creating several Environmental Protection Agency (EPA) Superfund sites.[65] As we have polluted and depleted our local water supply, we've bought thousands of acres of farmland in surrounding valleys to obtain their groundwater pumping rights for our use. We have spent over 4 billion dollars to construct, and 60 to 80 million dollars a year to operate, the Central Arizona Project (CAP), which diverts water from the Colorado River and pumps it 1,000 feet (304 m) uphill in an evaporation-prone canal over 300 miles (482 km) through the desert to reach our city.[66]

Projected population growth and increased water use is predicted to outstrip Tucson's "renewable" water supplies by 2025.[67] The Colorado River—designated America's Most Endangered River due to mounting problems with radioactive, human, and toxic waste in the water[68]—has been over-allocated to the point that the southernmost reaches of the river are severely diminished, crippling much of the Colorado River Delta's ecosystem and economy.[69] If the states upstream from Arizona on the Colorado River, and if Mexico below, take the full shares of Colorado River water granted them there will not be enough water

Fig. I.10A. The Santa Cruz River in Tucson, Arizona looking northeast from the base of A-Mountain in 1904. Note the braided running water, densely vegetated watershed, and cottonwood (white), willow, and mesquite trees growing in the floodplain. Credit: Arizona Historical Society/Tucson, AHS Photo # 24868

Fig. I.10B. The same stretch of the Santa Cruz River in 2004. Note the dry, channelized riverbed, and how much of the watershed has been replaced with paving, buildings, or bare, compacted earth. The cottonwood, willows, and most of the mesquites have disappeared with the depleted water table. The river park landscape drains most runoff to the riverbed and is dependent upon a drip irrigation system.

Fig. I.11A. A property losing 75% of its annual rainwater by directing the runoff from the impervious 75% section of its surface area (roof and driveway) to the street.

Fig. I.11B. The same property harvesting, and utilizing, 75% more of its annual rainwater by directing the runoff from the impervious roof and driveway to the permeable, sunken, and vegetated quarter of the site. Note: Sites with less permeable yard space and more impermeable hardscape will likely require a rainwater tank in addition to vegetated basins to handle the runoff.

left to meet Arizona's needs and fill the CAP canal in drought years.[70] In the meantime, we are importing about 2,000 pounds of salt with every acre-foot of CAP water we pump into the Tucson area.[71] That salt is an additional challenge and can become a contaminant for the already salt-prone alkaline soils of our desert environment.

We can make a shift. A 1/4-acre (0.1-ha) lot in Tucson receives about 67,000 gallons (253,621 liters) of salt-free rain in an average year. The average single-family residence in Tucson (assuming three people) uses about 120,000 gallons (454,248 liters) of water a year, and roughly half of that is for outdoor use. This suggests that most residential outdoor water needs could be met by harvesting the rainwater that falls on the property instead of pumping groundwater—especially if low water-use native plantings are integrated into the landscape design.[72] Reducing Tucson's consumption of groundwater and imported surface water is the key to shifting our city toward sustainable balance with our local water resources.

Tucson's average rainfall actually exceeds our current municipal water use (see box I.14), but most of this rainfall is drained away or lost to evaporation. Harvesting more of that rainwater, coupled with more conservation, brings us to our alternative path.

OUR PATH TO ABUNDANCE

By cycling the water we infiltrate, conserve, and clean within our lives and landscapes, we empower ourselves to do far more with far less. So, more is available for everyone, creating *abundance*. We enhance our own water resources and those of others, especially those downstream and downslope. Rather than *commodifying* fresh water or turning it into a limited-access commodity to be bought, sold, and hoarded, we *communify* it by working together to enhance our local water resources and manage their fair use and equal accessibility. As we enhance our natural resources (our "commons" (box I.12)) within our own lives and throughout our neighborhoods, the "community watershed" and the community resources are enhanced many times over!

This must occur in the more water consumptive government, commercial, agriculture, and industry sectors as well as at home, but at home is where it begins because every government official, teacher, student, businessperson, farmer, and industry worker lives in a home. If we realize the potential of water harvesting at home where it is easiest to do so, we can realize it elsewhere, because we will have learned from

Fig. I.12. Rain as our landscape's primary water source

direct experience, we will be motivated by our success, and we will be living the example we are trying to set.

The main goal of the abundance path is to use less water than nature renewably provides while consistently improving water quality, flow, and dependability and as a result, decreasing groundwater pumping and eliminating the need to import water. The first step is to strive to harvest more rainfall in our landscapes than we use from municipal water sources or private wells. This leads to a more sustainable hierarchy in the household and community management of our water resources in which:

- Rain is our primary water source (fig. I.12);
- Greywater is our secondary source;
- Municipal water or groundwater from private wells is strictly a supplemental source used *only* in times of need.

Think about how rainwater can be your primary water source, not just for your landscape, but for domestic needs as well. The following chapters and volumes show you how to do both.

So read on, harvest some rain, and grow abundance. Yet be warned: Once you start putting this information to work, every rainstorm could pump

Box I.14. Rain, Rain, Everywhere … Can We Stop To Think?

According to sustainable development consultant David Confer:

- Dividing the average annual precipitation falling on the surface area of Tucson, Arizona by its population (Pima Association of Governments 1998 data), then dividing again by 365 days per year, we find there is approximately 235 gallons per person (capita) per day (gpcd) (893 liters pcd) of rainwater compared to a 1998 total use of approximately 165 gpcd (627l pcd) delivered by municipal water companies for municipal and industrial uses.[73]

- As population densities increase, available rainfall gpcd will decrease. This would eventually become a problem if rain was the source of potable water, but it is not a problem for meeting the water needs of landscapes, because as population density increases, hardscape density also increases. In densely developed areas of Tucson, as much as 3/4 of the land is hardscape consisting of roofs, roads, parking lots, driveways, and sidewalks. If the rainwater that falls on hardscape was directed to and infiltrated within the remaining areas still available for vegetation, the 12 inches (304 mm) annual rainfall is concentrated fourfold, to almost 50 inches (1,270 mm) per year. This approximates the annual rainfall in Jacksonville, Florida (51 inches or 1,295 mm). This does not mean we can or should now plant the vegetation of Florida in Tucson—our rainfall patterns are more erratic, and our evapotranspiration rates are higher than Florida's. Rather, this illustrates the large volume of local rainwater we currently ignore or expel. The bulk of Tucson's rainfall is currently lost to runoff and evaporation. Nobody really knows the figure, but estimates of up to a 90% loss seem reasonable.[74] (See figure I.11.)

you with so much excitement and wonder that even if it's 3 A.M. when the clouds break you'll be running outside in your underwear to watch your landscape soaking up the water!

Good luck, and may your water-harvesting endeavors be all wet!

QUESTIONS AND ANSWERS ABOUT RAINWATER HARVESTING

Some short answers to questions I am often asked:

Doesn't harvesting rainwater deplete the water resources of those downstream?

The objective of most water harvesting is to create a "forested hillside" effect in your landscape and community that slows, not stops, the flow of water. This is because a forested hillside that quickly absorbs rainfall, then slowly and consistently releases it from its spongy soils over a period of weeks, months, or years, is much healthier for the stream below it than a denuded hillside that rapidly sheds water, creating sudden, sediment-laden flows downstream that dry up after just hours or days.

These water-harvesting strategies can reduce the amount of surface runoff traveling downstream, but they usually enhance the flow of wells, springs, streams, and rivers, because you are absorbing and cycling water in the landscape rather than quickly shedding it.

This book, and especially volumes 1 and 2, presents examples of rainwater-harvesting strategies that turned sporadic flow of streams and rivers into dependable year-round flow, of well levels that rose, and the creation of ephemeral springs. Water resources were enhanced for those downstream as well as for those doing the harvesting.

Doesn't rainwater harvesting mean you'll have standing water in which mosquitoes can breed?

Rainwater Harvesting for Drylands stresses small-scale strategies that can easily absorb harvested rain into the soil within a couple of hours after a storm. Mosquitoes need water standing for more than three days to complete their life cycle from eggs into adults.[75]

If you are harvesting rainwater in a cistern rather than the soil, you simply ensure that mosquitoes have no access. Techniques in this volume and still more in volume 3 (cisterns) show how to keep sunlight (which encourages algae and bacteria growth), insects, and critters out.

Do I need a tank or cistern to harvest rainwater?

Not necessarily; you can often easily and effectively harvest rainwater in the soil with simple earthworks presented throughout volume 2 (chapter 3 in this volume provides an overview of some of these). In fact, it is almost always less expensive to harvest the rain in the soil than in a tank, as the soil is already there, and has a far greater storage capacity.

Does it cost a lot to harvest rainwater?

If you harvest rainwater in the soil and you are doing the work yourself, it can be *free*. If you hire someone to create a rainwater-harvesting landscape for you, it should not cost much more than for a conventional landscape. No additional materials are required; you mainly just need to move more dirt.

If you are installing a cistern the cost will depend on the size and manufacture of the tank, how you plan to plumb it, and how you plan to use the water. Tanks range in price from $75 for a 55-gallon rain barrel on up. See volume 3 for more cistern options and the resources appendix (section I) in this volume.

A cost-effective approach is to develop a water budget. Figure how much water can reliably be obtained and sustained on site, then determine what means of harvesting best meets your needs. (See chapter 2 for more.) Passive water-harvesting earthworks (see volume 2) are typically 50 times cheaper than cisterns and can hold far more water. They can be used throughout a site's landscape, and are excellent for passively harvesting dirtier stormwater runoff, such as from streets and walkways. Use active systems (cisterns) to back up your passive system in drought. Only harvest a site's cleanest and most easily harvested water (typically from rooftops) in cisterns, so you'll get the most for your tank investment. And direct cistern overflow to water-harvesting earthworks.

Can I find a competent person to create my own water-harvesting landscape or system?

Rainwater harvesting is starting to become more commonplace, and while some areas have skilled designers and installation crews, many are lacking in such resources.

This book is meant to help more people become knowledgeable and skilled at water harvesting. Use it to guide a landscape architect if you hire one, or a crew doing the work on the ground. You will most likely need to do more supervising, but hey, you'll get a better job that way. Be careful whom you hire, check out the quality of their work, show them this book, and see how willing they are to work with you and new ideas.

Think about doing things yourself or with the help of friends. You'll learn a lot more and save money. As you learn by doing, you gain the skills that enable you to help others.

Are there rainwater-harvesting building codes?

Surface water laws vary around the country, so it's wise to check in with local authorities. In my area folks can harvest all the water that falls directly on their site, but there are restrictions on harvesting runoff within established waterways that pass through their site. Some arid counties have draconian laws prohibiting the harvest of rainwater runoff generated on site. In such instances, harvest the rain before it becomes runoff.

The International Building Code recommends avoiding the infiltration of water into soils within 10 feet of a building's foundation. If you have a basement, you may want to avoid infiltrating water within 20 feet of the foundation. Other than this, check with your local building inspector.

The International Residence Code (IRC) states no permit is needed for "Water tanks supported directly upon grade if the capacity does not exceed 5,000 gallons (18,927 liters) and the ratio of height to diameter or width does not exceed 2 to 1." Otherwise tanks need a permit.

The State of Ohio Department of Health and the State of Virginia Bureau of Sewage and Water Services regulate rainwater cistern systems, though in most counties and municipalities the unwritten code is "CYB"—"Cover Your Butt." That's exactly what the water-harvesting principles and strategies in this text strive to do for you. Follow them all and you won't have any mosquito, flooding, or drowning problems.

How can I learn more about water harvesting?

Do it! There is no better way than hands-on experience.

You can also look to the resource appendix in this book and also the resource page of my website www.HarvestingRainwater.com for a few of the organizations that offer courses in rainwater harvesting and tours of rainwater-harvesting sites. Many other publications and videos are also listed.

CHAPTER

The Man Who Farms Water and the Rainwater-Harvesting Guidelines

T his chapter is the core of the book and the heart of successful water harvesting. It describes eight guiding principles and three overriding ethics that are the foundation of how to conceptualize, design, and build integrated water-harvesting systems that generate multiple benefits. Use the principles and ethics as an integrated system, while thinking of them as a supportive mantra you can chant or a guiding song you can hum as you play in the rain and experiment with water harvesting.

Used together, these principles and ethics will greatly increase your chance of success, dramatically reduce mistakes, empower you to adapt various strategies to meet your site's specific needs, and allow you to expand the benefits of your work well beyond your site. Learn from doing, but don't go into it blind.

I begin with a story of the man whose life embodies the power of water harvesting, and who made it all click for me. …

THE MAN WHO FARMS RAINWATER

While traveling through southern Africa in the summer of 1995, I heard of a man who was farming water. I set out to find him and soon was packed into a colorful old bus roaring through the countryside of southern Zimbabwe. The scenery was beautiful, with rolling hills of yellow grass on red earth and small thickets of twisting, umbrella-like trees. Nine hours

later we arrived in Zimbabwe's driest region. We crested a pass of low-lying semi-desert vegetation. Below us spread a vast veldt prairie of undulating hills covered with dry grass and capped with barren outcroppings of granite. Trees were sparse. All was covered by a wonderful expanse of clear blue sky, reminding me of the open grasslands of southeastern Arizona. The bus crept down and stopped in Zvishavane, the small rural town where the water farmer lived.

In the morning, I hitched a ride with the local director of CARE International. She took me to a row of single-story houses. One of these was the simple office of the Zvishavane Water Resources Project. There on the porch sat the water farmer, reading a Bible.

As my ride came to a stop he sprung up with a huge smile and warm greetings. Here at last was Mr. Zephaniah Phiri Maseko. When he learned how far I had traveled, he burst into a wonderful laugh. He told me that lately visitors from all over the globe seemed to be dropping in once a week. Nonetheless, for him each was an unexpected surprise. Mr. Phiri jumped into the vehicle and we drove off over worn, eroded dirt roads toward his farm. An endless stream of poetic analogies, laughter, and stories began to pour from his mouth. The best story of all was his own.

In 1964, he was fired from his job on the railway for being politically active against the white-minority-led Rhodesian government. The government told him that he would never work again. Having to support a family of eight, Mr. Phiri turned to the only two things he had—an overgrazed and eroding 7.4-acre (3-hectare) family landholding, and the Bible. He used the Bible as a gardening manual and it inspired his future. Reading Genesis he saw that everything Adam and Eve needed was provided by the Garden of Eden. "So," thought Mr. Phiri, "I must create my own Garden of Eden." Yet he also realized that Adam and Eve had the Tigris and Euphrates Rivers in their region, while he didn't have even an ephemeral creek. "So," he thought, "I must also create my own rivers." He and his family have done both.

The family farm is on the north-northeast-facing slope of a hill providing good winter sun to the site since it is in the Southern Hemisphere. The top of the hill is a large exposed granite dome from which stormwater runoff once freely and erosively flowed. The average annual rainfall is just over 22 inches (570 mm). However, as Mr. Phiri points out, this average is based on extremes. Many years are drought years when the land is lucky to receive 12 inches (304 mm) of rain. When Mr. Phiri began, it was very difficult to grow crops successfully, let alone make a profit. There were frequent droughts and he had no money for deep wells, pumps, fuel, and other equipment needed for irrigating with groundwater.

Along with everyone else in the area, Mr. Phiri was dependent on the rains for water. Storms always brought him outside to observe how water flowed across his land. He noticed that soil moisture would linger longer in small depressions and upslope of rocks and plants, than in areas where sheet flow went unchecked (fig. 1.1). He realized he could mimic and enhance areas of his land where this was occurring, and he did so. He then spent ample time watching the effects of his work. Thus began his self-education and work in rainwater harvesting—his "water farming." Over the next 30 years, he created a sustainable system that now provides *all his water needs from rainfall alone* (fig. 1.2).

"You start catchment upstream and heal the young, before the old deep gullies downstream," says Mr. Phiri. Beginning at the top of the watershed, he built unmortared stone walls at random intervals on contour (along lines of equal elevation). These "check dam walls" slow or "check" the flow of storm runoff and disperse the water as it moves through winding paths between the stones. Runoff is then more easily managed because it never gets a chance to build up to more destructive volumes and velocities. Controlled runoff from the granite dome is then directed to unlined reservoirs just below.

Fig. 1.1. More water, soil, seeds, and life gather where their flow across the land is slowed (here by rocks on contour).

These reservoirs were built with nothing more than hand tools and the sweat of Mr. Phiri and his family. All work on the land was—and is—done on the human scale, so that it can be maintained on the human scale.

The larger of the two reservoirs Mr. Phiri calls his "immigration center." "It is here that I welcome the water to my farm and then direct it to where it will live in the soil," he laughs. The water is directed into the soil as quickly as possible. The reservoirs are located at the highest point in the landscape where soil begins to cover the granite bedrock. (See figure 1.3.)

Above the reservoirs the slope is steep with little soil. At and below the reservoir, the slope is gentle and soil has accumulated. "The soil," Mr. Phiri explains, "is like a tin. The tin should hold all water. Gullies and erosion are like holes in the tin that allow water and organic matter to escape. These must be plugged."

Mr. Phiri's "immigration center" is also a water gauge, for he knows that if it fills three times in a season, enough rain will have infiltrated the soil of his farm to support the bulk of his vegetation for two years. The reservoirs occasionally fill with sand carried in the runoff water. The sand is then used for mixing concrete, or for reinforcing the mass of the reservoir wall. Gravity brings this resource to Mr. Phiri free of charge.

Overflow from the smaller reservoir is directed via a short pipe to an aboveground ferrocement (steel-reinforced concrete) cistern that feeds the family's courtyard garden in dry spells. The family has another cistern, shaded and cooled by a lush food-producing passion vine (fig. 1.4). This cistern collects water from the roof of the house for potable use inside. Aside from these two cisterns, all water-harvesting structures on the farm directly infiltrate water into the soil where the water-harvesting potential

"As Mr. Phiri explains, 'I am digging fruition pits and swales to plant the water so that it can germinate elsewhere.'"

1. Granite dome
2. Unmortared stone walls
3. Reservoir
4. Fence with unmortared stone wall
5. Contour berm/terrace
6. Outdoor wash basin
7. Chickens and turkeys run freely in courtyard
8. Traditional round houses with thatched roofs
9. Main house with vine-covered cistern and ramada
10. Open ferro-cement cistern
11. Kraal—cattle and goats
12. Courtyard garden
13. Contour berm
14. Dirt road
15. Thatch grass and thick vegetation
16. Fruition pit in large diversion swale
17. Crops
18. Dense grasses
19. Well and hand pump
20. Donkey pump
21. Open hand-dug well
22. Reeds and sugar cane
23. Dense banana grove

(illustration by Silvia Rayces from a drawing by Brad Lancaster)

Fig. 1.2. Layout of Mr. Phiri's farm

Fig. 1.3. Mr. Phiri in his "immigration center" reservoir

Fig. 1.4. The family house, courtyard, cistern, and passion vine

Fig. 1.5. A loose-rock check dam that has healed a once-erosive gully

is the greatest. All greywater (used wash water) from an outdoor washbasin is drained to a covered, unmortared, stone-lined, shallow, underground cistern where the water is quickly percolated into the soil and made available to the roots of surrounding plants.

Across the farm's entire watershed from top to bottom, numerous water-harvesting structures act as nets that collect the flow of surface runoff and quickly infiltrate the water into the soil before it can evaporate. These include check dams (small unmortared stone walls placed within drainages perpendicular to the water's flow; see figure 1.5), vegetation planted on contour, terraces, berm 'n basins (dug out basins and earthen or vegetated berms laid out on contour), and infiltration basins (basins without berms). All these catch water that was once lost to a government-built drainage system.

Many years before, the government had built large drainage swales throughout the region. Unlike most water-harvesting swales or berm 'n basins, these ditches were not placed across the slopes on contour (to retain water), but instead were built so they'd drain water off the land. Vast amounts of unhindered monsoon runoff were caught by the drainage swale, carried away to a central drainage, and shot out to the distant floodplain. The erosion problem was decreased, but drought intensified because this area was being robbed of its sole source of water.

Mr. Phiri turned things around by digging a series of large "fruition pits" (basins about 12 feet long by 3 to 6 feet wide by 4 to 6 feet deep) in the bottoms of all the drainage swales on his land. Now when it rains the pits fill with water and the overflow successively fills one pit after another across his property. Long after rains stop, water remains in the fruition pits percolating into the soil. "You see," giggled Mr. Phiri, "my fruition pits are very fruitful." The fruit of the fruition pits takes the form of thatch grasses, fruit trees, and timber trees, which are planted in and around the pits. This vegetation provides building materials, cash crops, food, erosion control, shade, and windbreaks, all watered strictly by rain and the rising groundwater table underground. As Mr. Phiri explains, "I am digging fruition pits and swales to 'plant' the water so that it can germinate elsewhere." (See figure 1.6.)

"I have then taught the trees my system," continues Mr. Phiri. "They understand it and my language.

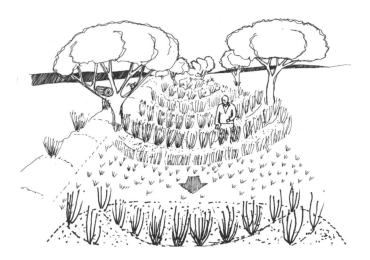

Fig. 1.6. Sketch of Mr. Phiri standing in a fruition pit full of thatch grass

Fig. 1.7. Mr. Phiri's tree nursery under the shelter of a mature tree

I put them here and tell them, 'Look, the water is there. Now, go and get it.'" A basin for holding water may be constructed around or beside the trees, but such earthworks are also placed further out from the trees so their roots are encouraged to stretch out and find still more water.

A diverse mix of open-pollinated crops such as basketry reeds, squash, corn, peppers, eggplant, tomatoes, lettuce, spinach, peas, garlic, onion, beans, passion fruit, mango, guava, and paw paws, along with such indigenous crops and trees as matobve, muchakata, munyii, and mutamba, are planted between the swales and contour berms. This diversity gives his family food security; if some crops fail due to drought, disease, or pests, others will survive. Rather than using hybrid and genetically modified (GMO) seed, Mr. Phiri uses open-pollinated varieties to create superior seed stock as he collects, selects, and plants seed grown in his garden from one year to the next. By propagating seed from plants that have prospered off the sporadic rainfall and unique growing conditions of *his* site, each season his seed becomes better suited to his land and climate. This seed saving is another form of water conservation, because Mr. Phiri adapts his seed to live off less water, instead of adapting his farm management to import more water.

Living fertilizer factories pepper the farm in the form of nitrogen-fixing plants. One example, the edible, leguminous pigeon pea, is also used for animal fodder and mulch. Mr. Phiri has found that soils amended with local organic matter and nitrogen-fixing plants infiltrate and hold water much better than those amended with synthetic fertilizers. As he says, "You apply fertilizer one year but not the next, and the plants die. Apply manure once and plant nitrogen-fixing plants, and the plants continue to do well year after year. Synthetically fertilized soil is bitter."

The abundant food and fruit Mr. Phiri produces is anything but bitter. He's been generous with his abundance, giving away a diverse array of trees to anyone who wants them. Unfortunately, as Mr. Phiri points out, the majority of the trees he gives away die when people don't implement rainwater-harvesting techniques before planting. "The land must harvest water to give to the trees, so before you plant trees you must plant water." Mr. Phiri propagates his trees in old rice and grain bags near one of three hand-dug wells near the bottom of his property (fig. 1.7).

The soil is Mr. Phiri's catchment tank, and it is vast. In times of drought, his distant neighbors' wells go dry, even those that are deeper than Mr. Phiri's. Yet as Mr. Phiri says, "My wells always have water into which I

Fig. 1.8. Mr. Phiri demonstrates how a donkey would power the pump in his lower fields.

Fig. 1.9. Mr. Phiri in his banana grove

Fig. 1.10. Mr. Phiri beside his largest aquaculture reservoir

can dip my fingers." This is due both to the particular hydrologic/geologic conditions of his site and because he is *putting far more water into the soil than he takes out.*

Except for one well, which is lined and has a hand pump for household water use, all are open and lined with unmortared stone. "These wells," explains Mr. Phiri, "are those of an unselfish man. The water comes and goes as it pleases, for you see, in my land it is everywhere." During severe drought, Mr. Phiri uses a donkey-driven pump to draw from these wells to water annual crops in nearby fields (fig. 1.8).

A lush wetland lies below the wells at the lowest point of Mr. Phiri's property. Here, three aquaculture reservoirs are surrounded by a vibrant soil-stabilizing grove of bananas (fig. 1.9), sugarcane, reeds, and grasses. The fish are harvested for food and their manure enriches the water used to irrigate the vegetation. The taller vegetation creates a windbreak around the ponds, reducing water loss to evaporation. The dense, lower-growing grasses filter incoming runoff water, as well as feed his cows when in calf. (See figure 1.10 of the largest reservoir.)

Mr. Phiri has created his Garden of Eden. The rain infiltrates his soil; the reservoirs and vegetation are where it "surfaces." This harvested rain creates the "rivers" of infiltrated moisture his Garden needed to succeed. After 30 years of work his farm continues to grow, and his methods are now starting to be appreciated.

For years Mr. Phiri was the object of scorn since he found himself in opposition to international aid and government programs that pushed groundwater extraction and export crops over rainwater harvesting and local food production and distribution. As a response Mr. Phiri created the Zvishavane Water Resources Project, a non-governmental organization that is spreading his techniques well beyond his site (see box 1.1). It is having a dramatic effect. He influenced CARE International in his region to the point that it shifted much of its work from giving away imported food, to helping people implement Mr. Phiri's methods of planting the rain and growing their own food.

When I asked Mr. Phiri about the three decades it took him to get his land and his vision to where it is today he answered, "It's a slow process, but that's *life.* Slowly implement these projects, and as you begin to

Box 1.1. Zvishavane Water Resources Project

If you'd like to support the great work of this grassroots project write to: Mr. Zephaniah Phiri Maseko, ZWRP, P.O. Box 118, Zvishavane, Zimbabwe.

To read more on Mr. Phiri see *The Water Harvester—Episodes from the Inspired Life of Zephaniah Phiri* by Mary Witoshynsky. Weaver Press, 2000. ISBN: 0-7974-2123-8.

Box 1.2. The Rainwater-Harvesting Principles

1. **Begin with long and thoughtful observation.**
2. **Start at the top—or highpoint—of your watershed and work your way down.**
3. **Start small and simple.**
4. **Spread and infiltrate the flow of water.**
5. **Always plan for an overflow route, and manage that overflow water as a resource.**
6. **Maximize living and organic groundcover.**
7. **Maximize beneficial relationships and efficiency by "stacking functions."**
8. **Continually reassess your system: the "feedback loop."**

Principles 2, 4, 5, and 6 are based on those developed and promoted by PELUM—the Participatory Ecological Land-Use Management association of east and southern Africa. Principles 1, 3, 7, and 8 are based on my own experiences and the insights gained from Mr. Zephaniah Phiri Maseko and other water harvesters.

rhyme with nature, soon other lives will start to rhyme with yours."

We then walked back up toward the house and stopped midway. Mr. Phiri's eyes were full of excitement and joy as he pointed across the fence. His neighbor was in the government's diversion swale, digging fruition pits on the adjoining property. "Look," cried Mr. Phiri, "he is starting to rhyme!"

My visit with Mr. Phiri made clear to me that we all have the choice and power to be either the problem or the solution. He told me of a local school where the teachers were striking and threatening to leave due to lack of water and harsh conditions in dusty, hot, wind-blown classrooms. Students were in no condition to learn, being malnourished without school meals and with little food at home. Mr. Phiri listened to the complaints of the teachers then asked them not to run from their problems. He told the teachers, "to look upon wherever they found themselves as home, to set their roots into the ground, and to work to nourish and improve their lives together." Mr. Phiri then made them an offer: If they would stay he would teach them and their students how to turn things around by harvesting the rainfall to grow food, shelter, and beauty. He also warned that if the teachers ran from the situation, they would take their problems with them. Half did leave. The other half stayed, set their roots, and worked with Mr. Phiri and the students. Together they turned the bleak school grounds into lush gardens where lunches are grown on-site and vegetation passively cools buildings and blocks the wind. There is no longer reason to strike or leave, but reason to celebrate.

Years later some of the teachers who left returned. With tears in their eyes they thanked Mr. Phiri for

being a man of his word. They also told him his prediction had come true. They had moved on to schools in new settlements in lush lands, but within a few years they had so misused and degraded the land that conditions became as bad as those from which they had run. Mr. Phiri responded by repeating his original offer. The teachers could go back to the schools in the new settlements and heal the scars.

Mr. Phiri turned to me with a huge smile and said, "Remember, children are our flowers; give them rain and they will grow and bloom."

THE EIGHT RAINWATER-HARVESTING PRINCIPLES

Mr. Phiri's story is a wonderful example of a successful, integrated rainwater-harvesting system (see chapter 4 for an urban example—that of my brother and myself). Keep in mind that the specific techniques used on his site are not applicable everywhere. There is no one standardized design for rainwater harvesting.

Every piece of land, the plants and animals upon it, and those who steward it, are unique. Each site must be approached with its own distinctive characteristics in mind. However, there are eight rainwater-harvesting principles that are applicable to all sites, and should always be followed. Each is valuable on its own, but you get the full benefit only if all are used together.

RAINWATER-HARVESTING PRINCIPLE ONE

Begin with Long and Thoughtful Observation

Mr. Phiri was not taught by experts or at schools. He learned from long and thoughtful observation of his land (fig. 1.11)—something everyone can do. When he began, his land was dry, eroded, and unproductive, but he was attentive to, and mimicked, the aspects of his land that were working—including rocks and plants found in informal "rows" perpendicular to the slope where they slowed the rainwater and infiltrated it into the soil. Mr. Phiri mimicked this by tucking his water-harvesting structures perpendicular to the slope around existing vegetation, placed at locations where they suited the needs of his family and land. He then spent ample time watching the effects of his work. As Mr. Phiri says, "I enjoy harvesting water. Really, you know, when the rains fall and I see water running, I am running! Sometimes you will find me being very wet!"

To observe your site, sit in the dust and dance in the rain, through all the seasons. Sit down, sit quietly, and turn yourself into a sponge. Listen with all your senses—sight, smell, touch, hearing, taste, and your feelings.

Observe all that is happening. Are there lush green areas where moisture naturally collects? Do you see bare spots where water and soil drain away? Is there running water? Is it polluted? If so, by what? Do trees grow straight, or are they bent—perhaps by strong prevailing winds? Is the soil underfoot washed-out and hard-packed, or soft with accumulated organic matter? Can you hear the songs of birds and insects? Some, such as house finches and dragonflies are never far from water. Where is the life? The resources? The erosion? Where is water coming from? Where is it going? How much water is here?

Relax and be aware. After you've taken plenty of time to observe, contemplate why things are as they are. Why is there erosion? Why does this plant grow here? Why is there more moisture there?

Try to understand the site as a whole, not as separate pieces. Imagine what would happen if you changed something. How would that alter the dynamics of the site's water flow, wildlife paths, prevailing winds, and solar exposure? How would things improve? How would they get worse?

If you listen, the land will tell you things you need to know, and what you need to investigate more deeply. Devoting time to observation and posing and answering questions are a good ways to get to know PLACE.

Once you connect to a place, it begins to show you its resources and challenges and helps guide your plans. Without understanding your site, you might install water-harvesting earthworks and plant fruit trees in an exposed, wind-dried area far from home where runoff is lacking and fruit-eating wildlife abounds. With better site understanding, you would locate these water-harvesting earthworks and trees where ample runoff, shelter from afternoon sun and prevailing winds, and a greywater-producing home are nearby, resulting in far more productive trees conveniently located for easy picking by people and less fruit-eating by wildlife. The outcome is determined by how well you understand your site, and how well you put things together.

It costs nothing to observe, think, and plan. You could pay dearly in the long run if you don't. Try to make your mistakes in your head or on paper before

Fig. 1.11. Long and thoughtful observation

doing anything on the land. Keep imagining different scenarios until you settle on the one you consider best.

RAINWATER-HARVESTING PRINCIPLE TWO

Start at the Top—or Highpoint—of Your Watershed and Work Your Way Down[1]

When you're feeling ready to create water-harvesting structures on your land, start at the "top"—or highest elevation—of your on-site watershed (fig. 1.12). You could also state this rule as "start at the beginning"—the beginning of the water's flow over your buildings and land.

To begin with, form a mental image of your *watershed*. A watershed, or *catchment area* as it is sometimes called, is the total area of a landscape draining or contributing water to a particular site or drainage. The watershed for an erosive rill cut on a bare slope might be no more than 20 square feet. The watershed for a river may be millions of acres covering mountains, hills, valleys, mesas, and drainages. Such large watersheds are made up of many small "subwatersheds." These subwatersheds are a patchwork quilt of small areas of land and buildings, often on the scale of a residential lot, a small parking lot, a commercial site, or a field. You will most likely be focused on the subwatersheds that comprise your home and workplace. These subwatersheds directly affect the larger community watershed, and if they are well managed have the potential to enhance the community watershed! Once you've identified the subwatershed of your site, you can begin to practice the art of *waterspread*, emphasizing the gentle harvesting, spreading, and infiltrating of water throughout a watershed rather than the rapid shedding or draining of water out of it.

Next, consider runoff. When more rain falls than surfaces can absorb, water pools and then begins to flow over roofs, roads, and soils on its way downslope. This surface flow of water is called *runoff*: water running off the land. Generally, the further you are downslope the greater the runoff volume will have accumulated. The steeper the slope the greater the speed of water you'll have to deal with. Our goal is

Box 1.3. Understanding Erosion and Water Flow Patterns

Erosion patterns, the presence of vegetation and wildlife, the size of deposited sediment, the smoothness of rocks, and other patterns give clues to how fast rainfall runoff flows through otherwise dry areas, how much water flows there, where it pools and where it moves. Appendix 1 delves into this more deeply.

Box 1.4. Additional Resources for Learning About Your Land

- Begin measuring and recording rainfall and other site-specific conditions.
- Talk to neighbors about their observations through the years.
- Seek out photos and writings at local historical societies.
- Obtain aerial photos from government mapping departments and private companies, or hire a small plane to fly you over your site and take photos. This gives you different perspectives of your site and surrounding areas, and documents land changes over time.

Fig. 1.12. Start at the top.

to turn this *runoff* into *soak-in*: water that no longer runs off the land, but infiltrates into the soil instead.

You may or may not have access to lands upslope of your property line, so begin water harvesting at the top of your "*watershed of influence*"—basically the sub-watershed composed of the area where you have the greatest say. This could mean a hilltop, the highpoint of your property line, the top of a cooperative uphill-neighbor's land, or the roof of your house.

If you begin to harvest water high in the watershed and work your way down, you'll make everything easier in the long run because:

• The volume of runoff you will be dealing with at any one time will be less than if you started lower in the watershed, and will be less likely to get out of control and become destructive. As a result, you can manage it better and construct a water-harvesting system in which most of the water will infiltrate before it runs off the land.

• You can use many modest-sized water-harvesting structures, each retaining an easily managed volume of water. Vegetation in modest-sized water-harvesting structures will get watered without getting flooded.

• The rain will infiltrate more evenly into the soil throughout the landscape, not just at the bottom.

• Water you harvest high in the watershed can be moved around the site more easily than water harvested low in the watershed. Gravity is a free and ever-present energy source that does not break down; use it to your advantage.

RAINWATER-HARVESTING PRINCIPLE THREE

Start Small and Simple

Small is beautiful, and perhaps more importantly when it comes to water harvesting, it is less expensive, easier, and more effective than starting big. Mr. Phiri and his family built everything by hand, spent almost nothing on materials, and did all the maintenance themselves. They could do this because everything was done on a human scale, and kept technically and mechanically simple to reduce the need for maintenance, and enable them to do that maintenance. (See figure 1.13.)

Small-scale trials of various techniques will quickly show you what works and what doesn't work on your

Fig. 1.13. Start small and simple, perhaps by planting a low-water-use native shade tree in a water-harvesting basin to shade the east or west side of your home.

unique site. You'll avoid large-scale mistakes. If a small-scale mistake is made, it will teach you, not break you. Starting small lets you and your friends do the work at your own pace, though of course you can hire folks to help with the work. Either way, don't start by creating an expensive and elaborate system that might not be right for your landscape, lifestyle, and means. Keep in mind that dozens, hundreds, or even thousands of tiny water harvesting "sponges" are usually far easier to create and far more effective than one big dam, because they capture more water and spread it more evenly throughout the land.

RAINWATER-HARVESTING PRINCIPLE FOUR

Spread and Infiltrate the Flow of Water[2]

Spread out the flow of water so it can *slow down* and *infiltrate* into the soil. Make water stroll, not run, through the landscape. This is the act of "waterspread" within the watershed.

Aside from one cistern holding water for a courtyard garden and another capturing roof runoff for household potable water, all Mr. Phiri's water-harvesting strategies direct the rain into the soil. He uses multiple techniques to spread harvested water over as much porous surface area as possible to give the water maximum potential to infiltrate *into* his land. Once it has infiltrated, water gently travels *through* the soil,

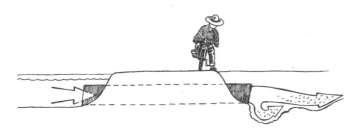

Fig. 1.14. Culvert acting as an erosive shotgun barrel. Note the undercut bed of the drainageway on the downstream side.

not destructively over it. As he says, "I plant water as I plant crops. So this farm is not just a grain plantation. It is really a water plantation."

In contrast, channelization can be compared to a shotgun barrel for water; it typically straightens and constricts water flow by sealing and smoothing the banks and sometimes the bed of a waterway, often with concrete. It's like the hardening of arteries in the body, and it's bad for the health of the system. Channelization increases the velocity of water flow through and downstream of the channelized area, reducing infiltration of water into the soil and sometimes deepening the channel.

A culvert (large pipe) placed in a drainage typically forces water flow through a smaller diameter orifice than the full width and depth of the natural drainage. In a large rain event, water backs up on the upstream side of the culvert, builds up pressure, and speeds through the culvert like it was the barrel of a shotgun. Resulting erosion can often be seen just downstream of the culverts. (See figure 1.14.) Yet, we can reduce erosion and enrich the landscape if we slow down the flow of water, spread it out, and allow it to infiltrate. In the second volume of *Rainwater Harvesting*, the chapter on check dams provides examples in drainages.

Figures 1.15A and 1.15B illustrate how a landscape can act as either a *drain* or a *net*, respectively. Water flows from the highest point or source of a watershed, to the bottom of the watershed or *sink* where the water and other resources leave the land for good (see figure 1.16 for Source/Sink).

The "drain" example of figure 1.15A shows water, soil, and organic matter quickly draining out of the system *causing* erosion and downstream flooding. Upstream areas are left dry while down-

---

Box 1.5. Small Dams Yield More Water Than Large Dams

A study by the Central Soil and Water Conservation Research and Training Institute in Dehra Dun, India, found that increasing the size of a dam's catchment from 2.47 acres (1 ha) to about 4.94 acres (2 ha) reduces water yield per hectare by as much as 20 percent.[3] As the Centre for Science and the Environment states, "In a drought-prone area where water is scarce, 10 tiny dams with a catchment of 1 ha each will collect much more water than one larger dam with a catchment of 10 ha."[4] The tiny dams don't need costly water distribution systems either, as they are already located throughout a watershed. Also, small dams displace far fewer people and cause less environmental damage than large dams.

In another example, tiny "dams" yield even more water than small dams. While studying 4,000 year old water-harvesting strategies in the Negev Desert, which enabled ancient people to provide food and water with a scant 4 inches (105 mm) of annual rainfall, Israeli scientist Michael Evenari found that *small watersheds harvest far more water than large watersheds*.

Summarizing Evenari's findings, the book *Making Water Everybody's Business* states, "While a 1 hectare watershed in the Negev yielded as much as 95 cubic meters of water per hectare per year, a 345 ha watershed yielded only 24 cubic meters of water/ha/year. In other words, as much as 75% of the water that could be collected [*in the larger watershed*] was lost [*to evaporation and the soil*]."[5] The loss was even higher during a drought year. According to Evenari "...during drought years with less than 2 inches (50 mm) of rainfall, watersheds larger than 123.5 acres (50 ha) will not produce any appreciable water yield, while small natural watersheds will yield 4,400–8,800 gallons (20–40 cubic meters) per hectare, and microcatchments smaller than 0.24 of an acre (0.1 hectare) [*will yield*] as much as 17,597–21,997 gallons (80–100 cubic meters) per hectare."[6]

---

stream areas require expensive stormwater management. The system degenerates—or breaks down over time.

The "net" example (fig. 1.15B) shows the same area with the landscape altered to capture, slow, and

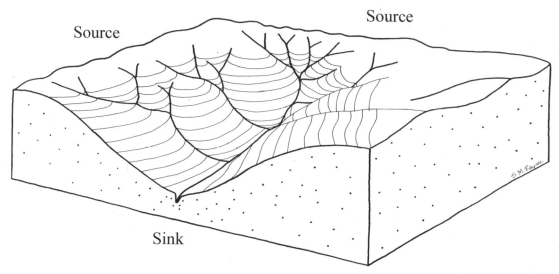

Source

Source

Sink

Fig. 1.15A. A bare landscape acting as a drain. The water's Source is at the top; its Sink is at a lower elevation.

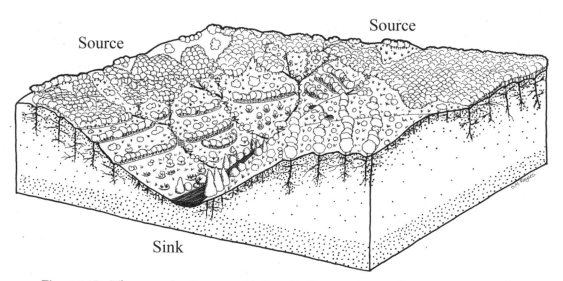

Source

Source

Sink

Fig. 1.15B. The same landscape with "nets" of vegetation and water-harvesting earthworks slowing and infiltrating the water.

spread the flow of water, soil, and organic matter, over and through the entire landscape. This *reduces* erosion, flooding, drought, and monetary costs of storm drain infrastructure while *improving* soil fertility, water infiltration, vegetative production, and ecosystem stability—by growing shade, food, shelter, wildlife habitat, and erosion control. This system starts to regenerate—or build and take care of itself.

The goal in water harvesting is to create a series of "nets" across our watershed. Like Mr. Phiri, we should direct runoff *into* the soil by spreading and sinking its flow. Still, there will always be storms so big that more water will flow across the site than the land or tanks can retain, leading us to the next principle.

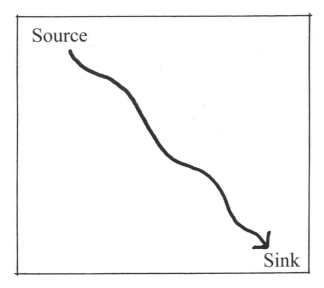

Fig. 1.16A. Source and Sink. A fairly quick and linear downward flow drains the landscape.

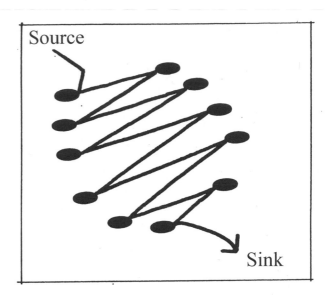

Fig. 1.16B. Source and Sink. The zig-zag increases the time of flow, distance traveled, and ground infiltration from Source to Sink.

RAINWATER-HARVESTING PRINCIPLE FIVE

Always Plan for an Overflow Route, and Manage That Overflow Water as Resource[7]

Overflow should not be treated as a problem or a waste. Instead, design the overflow route so that surplus water becomes a *resource* (fig. 1.17). Mr. Phiri converted the government-built drainage swales cut across his land into a water-harvesting project by digging "fruition pits"—or stepped infiltration basins—within the bottom of the large swales. Any excess water overflows from one fruition pit to the next and ultimately drains away down the big swales. All of Mr. Phiri's water-harvesting structures have planned overflow routes. In huge storms extra runoff is directed from one harvesting structure to the next until it reaches the bottom of his site where it is released onto the natural floodplain below.

No matter how well you design your system *always* plan for overflow in very large storm events. Overflow spillways should be stabilized using heavy tightly packed rock, or well-rooted vegetation so they hold up to large flows. Overflow from tanks and cisterns must be safely routed out of the tank and away from the tank's foundation. Overflow should

be directed to a useful location such as a vegetated infiltration basin that passively irrigates a native shade tree that in turn shades the tank, provides food, and creates wildlife habitat. The need to manage overflow applies to all cisterns and water-harvesting earthworks. As the Boy Scouts say, "Be Prepared." Make sure that when your system overflows, it overflows where you want it to, and in a controlled manner.

Be sure your site has a final overflow outlet at the bottom of your watershed. Ideally this would direct water into a natural vegetated wash or creek, but in

Fig. 1.17. Cistern overflow water directed to, filling, and then overflowing one earthwork to fill another and another

Fig. 1.18. Planting vegetative groundcover, spreading organic mulch, and planting seed to help permeate, protect, and build soil with roots, leafy cover, mulch, and accumulating leaf drop

Fig. 1.19. A water-harvesting system providing multiple functions of water, passive cooling with shade, stormwater control within earthworks, wildlife habitat, and food production.

the urban environment you may have to settle for a street or storm drain.

RAINWATER-HARVESTING PRINCIPLE SIX

Maximize Living and Organic Groundcover[8]

Rather than infiltrating, water often flows off flat or mounded bare dirt surfaces, or pools for days inside bowl-shaped surfaces and evaporates or supports mosquito breeding. This is because bare dirt is prone to compaction and the surface tends to seal up, both of which reduce the ability of rainwater to infiltrate below the surface. In contrast, covering dirt with organic groundcover such as mulch and plantings increases infiltration into the soil. Growing plants set down roots and drop leaves to generate mulch. Earthworms and other soil life convert the leaf drop into more soil, riddled with their holes. I've found the roots of these plants, coupled with surface mulch and the associated soil life, grow to create a living sponge that can more than double stormwater infiltration rates in previously bare basins, reducing evaporation, runoff, erosion, and mosquito breeding! (See figure 1.18.)

Mr. Phiri's site is a living vegetation-covered welcome mat that helps water infiltrate into the soil and pumps soil moisture back to the surface through

roots. The vegetation literally brings harvested water to "fruition," transforming it into fruits, vegetables, and grains for people, livestock, and wildlife; shade and shelter for home and fields; a dense mat of roots and leaves to stabilize spillways and control erosion; lumber and thatch for building; fiber for clothes; medicinal herbs; windbreaks that reduce evaporation and evapotranspiration; and leaf drop that breaks down and fertilizes the soil.

Native vegetation—indigenous plants found within 25 miles (8 km) of your site and within an elevation range of 500 feet (152 m) above or below your site—is generally best adapted to local rainfall patterns and growing conditions, and these plants often make great groundcovers.

RAINWATER-HARVESTING PRINCIPLE SEVEN

Maximize Beneficial Relationships and Efficiency by "Stacking Functions"

Mr. Phiri looks well beyond water infiltration and strives to improve his *whole* site, not just one aspect of it. He does this by designing and placing his water-harvesting structures in relationship to the overall landscape so they perform multiple beneficial functions—he is "stacking functions." By stacking functions, Mr. Phiri gets far more efficiency and productivity for the same amount of effort. The vegetation selected to harvest rainwater also produces food, medicine, fiber, shelter, wildlife habitat, and windbreaks. These windbreaks reduce evaporation of water from fields and ponds. Fish raised in these ponds feed the family and fertilize the water used in the fields. Contour berms create raised footpaths. Check dams stabilize path and road crossings over drainages. (See figure 1.19.)

Often, existing strategies designed to perform one primary function can be adapted to perform additional functions. For example, the government's drainage swales were originally designed only to reduce erosion and flooding, which they did, but they also drained away the area's sole source of water—an irreplaceable resource. Mr. Phiri added fruition pits to harvest water within the swales and lined his fruition pits with multi-use plants, creating windbreaks, stabilizing the pits, and generating self-seeding crops that flourish on passively harvested water.

Each site has its own unique potential for stacking functions. For example, when designing rainwater cisterns into a site, they can double as privacy walls, pillars supporting porches, property fences, retaining walls, afternoon sunscreens, and more. Get the Domino Theory working for you. You know you're doing well when you devise a strategy to solve one problem that simultaneously solves many other problems and creates more resources.

RAINWATER-HARVESTING PRINCIPLE EIGHT

Continually Reassess Your System: The "Feedback Loop"

Continual reassessment is the key to long-term maintenance of a water-harvesting system (fig. 1.20).

Mr. Phiri had a great idea: Grow water-harvesting structures by placing plants on contour. He "stacked functions" by seeking out plant species that produced

Fig. 1.20. Long and thoughtful observation again. How is the land responding to your work? What still needs to be addressed?

crops as they harvested water and reduced erosion. He quickly settled on hardy sisal plants (*Agave spp.*) that use little water, require almost no maintenance, produce large amounts of biomass to hold back water and soil, and produced fiber to use on site or sell.

Mr. Phiri thought long and hard about this strategy. He began high in his watershed where the sisal contour berms helped spread runoff and infiltrate it into the soil. He built rock-stabilized spillways for overflow. His system quickly maximized groundcover, harvested water, stabilized soil, and produced sisal fibers. The only thing he forgot to do was to start small.

Sisal plants covered his land and slowed runoff. Everything was great until winter arrived in a drought year. Grass was sparse, and plants appeared yellow and dead in their dormancy, but the evergreen sisal stood out verdant and lush: Mr Phiri's livestock went right for it. The long, strong fibers of the plants bound up in the intestines of the animals, killing them. Mr. Phiri was devastated. He had not foreseen these consequences. There are always consequences that we cannot or do not predict. Mr. Phiri subsequently spent

Fig. 1.21. Mr. Phiri beside his remaining stand of sisal planted on contour

many hard days removing all but one small stand of sisal. He left this stand as a reminder and teacher—and keeps the livestock away from it (fig. 1.21).

Following *all* the principles *together* can decrease your mistakes and increase your chances of success. No matter how good a plan or design is, maintenance and adaptation will be required over time. When the design is well thought out in the first place, these changes are likely to be minor. Mr. Phiri finds himself reinforcing spillways, maintaining berms and swales, and pruning vegetation for livestock forage and mulch. Sometimes, as with the sisal, he needs to change or alter some of his strategies. After the sisal mistake, Mr. Phiri did not abandon his idea of planting vegetation as water-harvesting structures—he was just more careful about plant selection. Now he starts with smaller plantings to see their effects before he expands to larger areas.

As with Mr. Phiri's site, all landscapes are continually evolving and we need to continually work *with* them. Go back frequently and observe how your site is performing, repair elements if needed, and see if there are ways you could improve on your site plan and techniques. We cannot escape the need for maintenance, but we can reduce the need for excessive maintenance by following the eight water-harvesting principles. Balanced maintenance should not be feared or neglected; it is an opportunity to learn and to improve.

WATER-HARVESTING ETHICS

Mr. Phiri's site and life provide a wonderful example of embodying water-harvesting principles within an integrated system. They also embody an ethical basis that further increases the benefits of his work. The three ethics of permaculture[9] described below are realized in Mr. Phiri's work, and are important guides to me in making decisions about water-harvesting and integrated-system design.

1. The CARE OF THE EARTH[10] ethic reminds us to care for all things living and nonliving, including soil, water, air, plants, animals, and entire ecosystems. As Bill Mollison states in *An Introduction to Permaculture*, "It implies harmless and rehabilitive activities, active conservation, ethical and frugal use of resources, and 'right livelihood' (working for useful and beneficial systems)."[11]

2. CARE OF PEOPLE[12] directs us to strive to meet our basic needs for air, water, food, shelter, education, fulfilling employment, and amiable human contact in ways that do not hamper or prevent others from doing the same. We do not exploit or disregard others for our own gain. Nor do we destroy the environment that supports us all. Instead, we sustain a basic quality of life that improves our environment while enabling others to do the same.

3. REINVESTMENT OF SURPLUS TIME, MONEY, AND ENERGY[13] to achieve the aims of earth and people care encourages us to extend our influence and surplus energies to help others attain the ethics in their own life and work. This helps us all because it strengthens the greater communities in which we all live.

Mr. Phiri embodies these ethics: He improves his land, the earth, and his community by working with local resources so his land and community can sustainably regenerate more resources. He eschews synthetic fertilizers and clear-cutting that provide short-term gains but pollute and weaken the land in the long run. He practices infusion rather than extraction. He gives his land more than he takes—in the form of water. He gives his community more than he takes—in the form of information, trees, and water. He empowers others to do the same. The Zvishavane Water Project was formed by Mr. Phiri to contribute surplus time, energy, and money to spread these ideas. He teaches people how to harvest rain, improve soil, grow food, and build community. And he learned it all from living it.

By following these eight principles and the "care" ethics above, you can thoughtfully and effectively practice water harvesting and create the best techniques and strategies for your unique situation. Use the principles and ethics as a checklist of guidelines while you assess your site, imagine what water-harvesting strategies would work best and where,

and as you implement your ideas. As long as all the guidelines are met you'll be on the path to abundance.

You now have the tools to conceptualize and plan an integrated rainwater-harvesting system. Read on to find out how much rain you have to harvest.

Box 1.6.
Additional Permaculture Resources

I invite you to look into permaculture as a tool to improve your rainwater harvesting efforts, and to integrate sustainable strategies into your life. While I offer a set of principles geared specifically to water harvesting, permaculture literature provides principles that apply to all aspects of our surroundings. I encourage you to pursue this exciting and empowering material. See appendix 6, section E.

Assessing Your Site's Water Resources

But especially as I drink the last of my water, I believe that we are subjects of the planet's hydrologic process,
too proud to write ourselves into textbooks along with clouds, rivers, and morning dew.
When I walk cross-country, I am nothing but the beast carrying water to its next stop.

—Craig Childs, *The Secret Knowledge of Water*

It's important to begin harvesting water knowing *how your site fits into larger water flows* and knowing *how much water there is to harvest.*

This chapter builds on the principle of long and thoughtful observation by beginning with a description of the hydrologic cycle. This will help you understand your site's water flow in the context of global hydrologic patterns and interconnections. Next the chapter moves to local watersheds and subwatersheds, where small-scale portions of the hydrologic cycle occur. You will learn how to determine the boundaries of your site's subwatershed, where to concentrate your water-harvesting efforts, how to calculate rainfall and runoff volumes that affect your site, and how greywater can add to your site's harvestable water resources. The chapter ends with the story of an Arizona couple striving to live within their site's rainwater budget. Appendix 3 "Calculations" and appendix 5 "Worksheets" are intended to be adjuncts to this chapter.

THE HYDROLOGIC CYCLE—OUR EARTH'S CIRCULATORY SYSTEM

The following is primarily drawn with permission from water-harvester Ben Haggard's great little book, *Drylands Watershed Restoration*:

Of the world's total water, a small percentage is fresh water. The majority of that is tied up as ice in polar ice caps and glaciers. The remainder is continually recycled in order to support the world's living systems. This recycling is known as the hydrologic cycle.

Water is evaporated from the oceans and precipitated as rain or snow over the continents. This water is absorbed by plants and evapotranspired back into the air. This pumping of water back into the air by plants accounts for much of our atmospheric water. This water forms clouds and rains again … Forests play an important role in maintaining [and retaining] rain in the landscape.

Raindrops form around ice crystals in clouds. These ice crystals require a nucleus for their formation. Dust, tiny bits of leaf, and bacteria are among the particles that initiate rain. A number of natural systems [such as forests] encourage rain by giving off columns of tiny particles that seed the clouds causing drops to form.[1]

Raindrops are soaked up by the living sponges of forests, prairies, and desert thornscrub. These, along with their associated leaf drop, topsoil, and the cavities created by burrowing animals, help hold onto that water and slowly release it.

If forests, grasslands, and other rain seeders and sponges are removed from a landscape, the landscape begins to dry. Rain can become less common and vegetation has trouble reestablishing. Rivers and streams become dry.

Ben Haggard continues:

Rivers and streams generally flow throughout the year; in spite of the fact that rain is a localized and fairly infrequent event in arid settings. Even in rainy climates, rain occurs a relatively small percentage of the time. Rivers have a sustained flow because most of the water is actually stored in the soil where it slowly releases into the drainage. In disturbed watersheds, this slow and sustained release is disrupted. Water runs rapidly off the ground's surface rather than soaking into the ground. This process creates floods followed by drought. To repair such a watershed, infiltration of the water into the ground must be increased.

Living systems create complex interactions with water. Water falls as rain. Trees intercept this water, directing it into the ground where a layer of organic material deposited by the trees absorbs and holds it. Some of the water flows slowly through the ground where it supports the growth of forests and the sustained flow of rivers. Rivers act as transportation networks, allowing nutrients from the forests to wash downstream and fish and other animals to swim upstream, importing phosphate and other minerals into the forests. Water in the landscape also attracts and supports wildlife, the active planters, fertilizers, and maintainers of the forest. The forests breathe water back into the air, where it condenses around particles also released by the forests. The clouds form and the entire process repeats itself."[2] (*Reprinted with permission from* Dryland Watershed Restoration—Introductory Workshop Activities *by Ben Haggard, copyright 1994, Center for the Study of Community.*)

Each of us depends on and is a part of the hydrologic cycle. As water moves through the global cycle, so it moves through the watersheds of our communities, the subwatersheds of our individual sites, and our own bodies, which are over 70% water. We can slow, cycle, and enhance that flow as we improve our lives and community by harvesting rainwater. First, we need to identify and thoughtfully observe our watersheds.

WATERSHEDS AND SUBWATERSHEDS— DETERMINING YOUR PIECE OF THE HYDROLOGIC CYCLE

A *watershed* is the total area of land from which water, sediments, and dissolved materials flow by gravity to a particular end point. At the largest scale this endpoint might be a river, lake, or ocean. A watershed is a geographic entity clearly defined by high points or ridgelines that split the flow of water, creating the boundaries of each watershed. Watersheds are made up of many smaller *subwatersheds*, each defined by lower elevation ridgelines that further split the flow of water and direct portions to particular endpoints. These subwatersheds are made up of a collection of still smaller subwatersheds. If you trace the boundaries of all these subwatersheds, the pattern you'll see looks like pieces of jigsaw puzzle forming an interconnected whole. The terms "watershed" and "subwatershed" are relative terms that can refer to a variety of scales of water drainage areas.

On a large scale, your land will almost surely be part of a regional watershed that drains thousands of square miles of land, creating streams and rivers. My Tucson home is a part of the Santa Cruz River watershed, which covers approximately 8,600 square miles in southern Arizona and northern Mexico. Within this regional watershed, water drains toward me from upslope areas, away from me to downslope areas, and ultimately flows to the Santa Cruz River. The Santa Cruz River watershed is in turn a subwatershed of the larger Gila River watershed, which is in turn a subwatershed of the still larger Colorado River watershed.

The land area of a city flowing toward one regional watershed consists of smaller subwatersheds throughout the city, broken up into many smaller neighborhood-

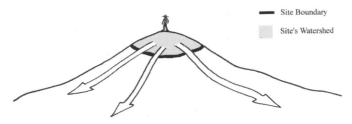

Fig. 2.1A. The top of the watershed

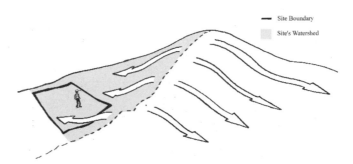

Fig. 2.1B. Ridge of hill defining watersheds

sized subwatersheds, made up of property-sized subwatersheds, made up of still smaller subwatersheds consisting of residents' roofs, yards, patios, and driveways. Each small urban-scale subwatershed directs flowing water toward a different urban endpoint. Urban landforms, buildings, and parking lots act as the dividing "ridges" between these tiny watersheds. Pitched roofs divide the flow of water between front and back yards. Parking lots act as gradually sloped fields. Roads act as linear ridges if raised, or as drainageways if built as a lower element in the landscape.

If your site is at the very top of a hill it is also at the top of a watershed, because all runoff water will be draining off the hill away from your site (fig. 2.1A). If your site is at the bottom of the hill, your site's watershed will be that part of the hill's slope that drains or sheds water toward your property. Most likely, the water on the other side of the hill will drain to a different endpoint, so it will be part of a different watershed (fig. 2.1B). But, if runoff from the other side of the hill does eventually drain toward your site—perhaps via an arroyo or wash curving around the hill—then it too is part of your site's watershed. In addition to this local hill, there may be other areas of land that drain toward your site. If so, the watershed affecting your site is even larger (fig. 2.1C).

IDENTIFY YOUR SITE'S WATERSHED AND OBSERVE ITS WATER FLOW

To assess your site's water resources, first define the boundaries of your property and the watershed directly affecting your site. A topographic map will give you a general idea of your watershed's "ridgelines"—the tops of slopes that determine if water is

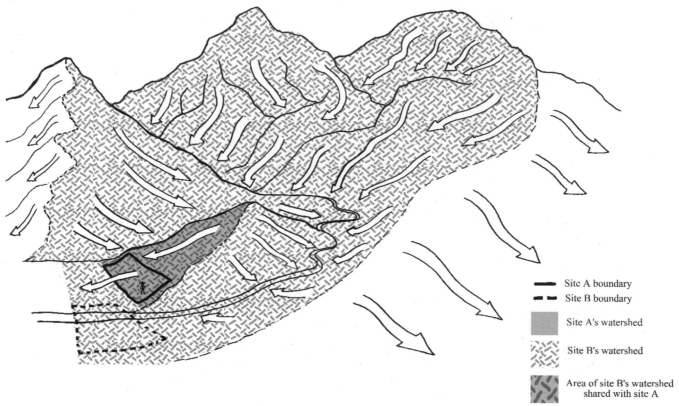

Fig. 2.1C. Watersheds and subwatersheds, the larger picture

Legend:
- — Site A boundary
- ▪ ▪ Site B boundary
- Site A's watershed
- Site B's watershed
- Area of site B's watershed shared with site A

flowing toward or away from your location (see box 2.2). You can walk your land in the rain to see which way water runs to help understand land slope and the extent of the watershed draining to your site. Erosion patterns can clue you in to flow patterns when it's dry (see appendix 1). If runoff flows across your land, pay particular attention to what direction it comes from, its volume, and the surfaces it flows over. Potential contaminants—such as oil from streets and pesticides from yards and fields—might be picked up and carried in this runoff water. (See figure 2.2 as to how you might conceptualize your site's runoff and runon.)

CREATE A SITE PLAN AND MAP YOUR OBSERVATIONS

Creating a site plan helps you see and make use of site resources and challenges; integrate your water-harvesting system with the rest of your site (see chapter 4);

and place and size water-harvesting earthworks, vegetation, and tanks appropriately.

You can use your own paper (perhaps you can create your own water-harvesting journal) or use the worksheets and grid paper provided in appendix 5.

Leave wide margins around the outside of the paper, and draw your property's boundaries "to-scale" inside these margins. If you choose a scale of 1/8 inch = 1 foot, a measured distance of 1 foot on your site will be drawn on your plan as 1/8 inch. Use the wide margins to map the locations where resources—such as runoff from your neighbor's yard—flow on, off, or alongside your site. Draw buildings, driveways, patios, existing vegetation, natural waterways, underground and above-ground utility lines (to avoid damaging them and yourself), and other important elements of your site to-scale on the plan. Make multiple copies of your basic site plan on which to draw a number of drafts of your observations and ideas. (Figure 2.3 is a sample site map.)

Box 2.3. Calculating Rainfall Volumes

To calculate the volume of rainfall in *cubic feet* that falls in an average year on a specific *catchment area*, such as your yard, neighborhood, or other subwatershed (see box 2.4 for roofs):

CATCHMENT AREA (in square feet) multiplied by the AVERAGE ANNUAL rainfall
(in feet) equals the TOTAL RAINWATER FALLING ON THAT CATCHMENT
IN AN AVERAGE YEAR (in cubic feet)

(or)

CATCHMENT AREA (ft^2) $\times$ AVG RAINFALL (ft) = TOTAL RAINWATER (ft^3)

If you normally measure annual rainfall in inches, simply divide inches of rain by 12 to get annual rainfall in feet. For example, folks in Phoenix, Arizona get about 7 inches of annual rainfall, so they would divide 7 by 12 to get 0.58 foot of annual rain.

Once you get your answer in cubic feet of annual average rainfall, convert cubic feet to gallons by multiplying your cubic foot figure by 7.48 gallons per cubic foot. The whole calculation looks like this:

CATCHMENT AREA (ft^2) $\times$ RAINFALL (ft) $\times$ 7.48 gal/ft^3 = TOTAL RAINWATER (gal)

For example, if you want to calculate how much rainwater in gallons falls on your 55 foot by 80 foot (4,400 square feet) lot in an normal year where annual rainfall averages 12 inches the calculation would look like this:

4,400 square foot catchment area $\times$ 1 foot of average annual rainfall $\times$ 7.48 gallons
per cubic foot = 32,912 gallons of rain falling on the site in an average year

To calculate the volume of rainfall falling on a specific catchment area in liters:

CATCHMENT AREA (in square meters) $\times$ AVERAGE ANNUAL RAINFALL
(in millimeters) = TOTAL RAINWATER FALLING ON A CATCHMENT
AREA IN AN AVERAGE YEAR (in liters)

To calculate the volume of rainfall on a specific catchment for a given rain event:

Use the calculations above, but enter the amount of
"rainfall from a given rain" in place of "average annual rainfall."

Note: Appendix 3 "Calculations" provides more detailed information on conversions, constants, and calculations for water harvesting.

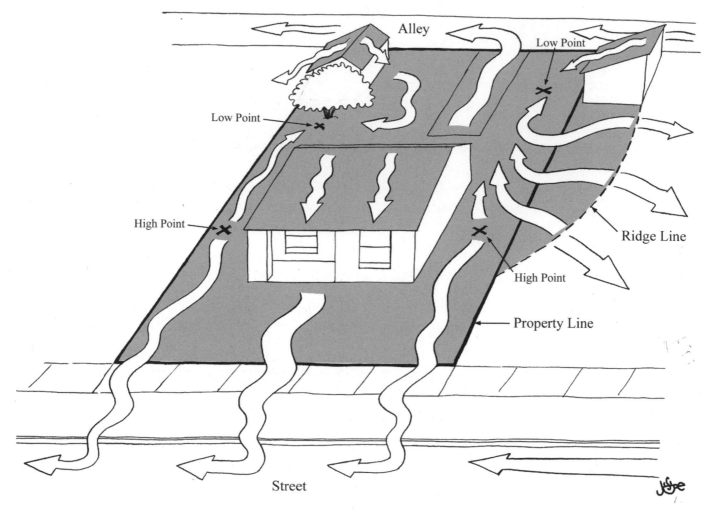

Fig. 2.2. An urban home watershed with arrows depicting runoff flow

CALCULATE YOUR SITE'S RAINFALL VOLUME

Once you've defined and mapped the boundaries of your site, use the calculations in box 2.3 to determine average volume of rain falling on your site each year. This is the "income" side of your "water budget." (Again see figure 2.3.)

MAP YOUR SITE'S CATCHMENT SURFACES AND CALCULATE THEIR RUNOFF VOLUMES

Roofs, paved surfaces such as driveways and patios, and compacted earth surfaces such as paths are useful catchment surfaces from which to harvest rainfall. Indicate on your plan any catchment surfaces that drain water *off* your site (for example, a driveway sloping toward the street) and *subtract* this lost runoff volume from your site's calculated average annual rainwater resources. (You can devise strategies to recapture that lost runoff later.)

Indicate on your plan any catchment surfaces draining water *onto* your site from off-site (for example, runoff from your neighbor's yard that drains into your yard). *Add* this bonus runoff (or "*runon*") volume to your site's calculated average annual rainwater resources. See box 2.4 for instructions on calculating runoff volumes. See figure 2.4 for the example site map with runoff volumes and runoff coefficients.

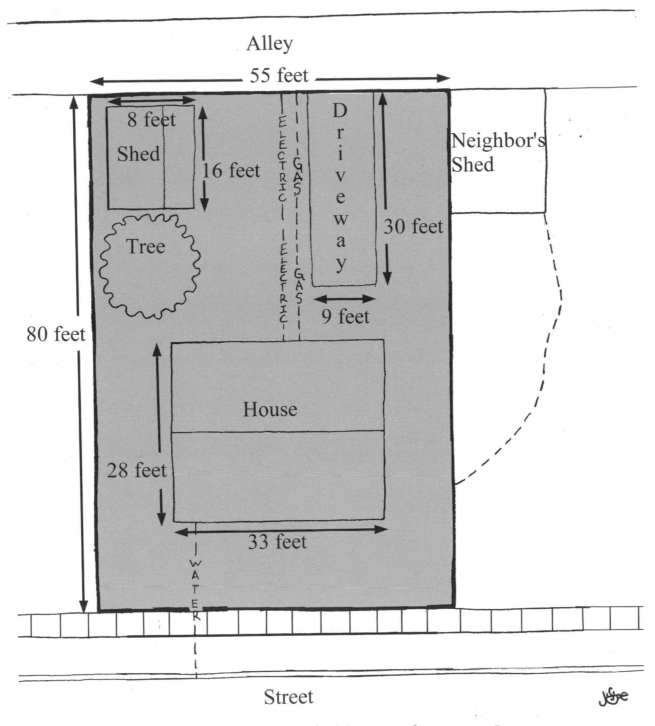

Fig. 2.3. Site map (overhead/plan view) of a 4,400 square foot property. In an average year of 12 inches of precipitation the site receives 32,912 gallons of rainfall "income."

Box 2.4. Calculating Runoff Volumes

You can get a ballpark estimate of runoff volume from any sloped surface by multiplying the volume of rain that falls on that surface by its "runoff coefficient"—the average percentage of rainwater that runs off that type of surface. For example, a rooftop with a runoff coefficient of 0.95 estimates that 95% of the rain falling on that roof will run off.

The runoff coefficient for any given surface depends on what the surface is composed of. Rainfall intensity also affects the coefficient: the higher the rainfall intensity, the higher the runoff coefficient. Ranges and averages of various runoff coefficients I use in the southwest U.S. are as follows:

• A roof or impervious paving: 0.80–0.95
• Sonoran Desert uplands (healthy indigenous landscape): range 0.20–0.70, average 0.30–0.50
• Bare earth: range 0.20–0.75, average 0.35–0.55
• Grass/lawn: range 0.05–0.35, average 0.10–0.25
• For gravel use the coefficient of the ground below the gravel.

The runoff coefficient for earthen surfaces is greatly influenced by soil type and vegetation density. Large-grained porous sandy soils tend to have lower runoff coefficients while fine-grained clayey soils allow less water to infiltrate and therefore have higher runoff coefficients. Whatever your soil type, the more vegetation the better, since plants enable more water to infiltrate the soil.

CALCULATING ROOF RUNOFF: AN EXAMPLE

Determine the size of a roof catchment by measuring only the outside dimensions—or "footprint"—of the roof's edge (if your house has a roof with overhangs the roof's footprint will be larger than the building's footprint). Ignore the roof slope; no more rain falls on a peaked roof than falls on a flat roof with the same footprint. (See figure 2.5.)

To calculate the runoff in gallons from a metal roof's 28 foot × 33 foot "footprint" (924 square feet) in a climate averaging 12 inches of rain a year:

924 square feet roof × 1 foot of average annual rainfall × 7.48 gallons per cubic foot
= 6,911 gallons of rain falling on the roof in an average year.

$$924 \text{ ft}^2 \times 1 \text{ ft} \times 7.48 \text{ gal/ft}^3 = 6{,}911 \text{ gallons/average year}$$

Multiply the above figure by the roof surface's runoff coefficient 0.95*:
6,911 gallons × 0.95 = 6,565 gallons of rain running off the roof in an average year.

Note*: 5 to 20% of runoff from impervious catchment surfaces such as roofs can be lost due to evaporation, wind, overflow of gutters, and minor infiltration into the surface itself. In volume 3, the chapter on cistern components includes a table for runoff coefficients specific to roof type.

CALCULATING YARD RUNOFF: AN EXAMPLE

Let's say we are on a site receiving 18 inches of rain in an average year, and the neighbor has about a 25 foot by 12 foot bare section of his yard that drains onto our example property. The soil is clayey and compacted.

Determine the available rainwater running off that section of the neighbor's yard onto our land by multiplying its catchment area (300 square feet) by the average annual rainfall in feet (1.5) by 7.48 (to convert the answer to gallons):

$$\text{CATCHMENT AREA (ft}^2\text{)} \times \text{RAINFALL (ft)} \times 7.48 \text{ gal/ft}^3 = \text{TOTAL RAINWATER (gal)}$$

300 × 1.5 × 7.48 = 3,366 gallons of rain falling
on that section of the neighbor's yard in an average year.

Multiply that figure by the soil surface's runoff coefficient of 0.60:

3,366 × 0.60 = 2,019 gallons annually running off the neighbor's compacted
yard into ours. Add that to our site's annual rainwater budget.

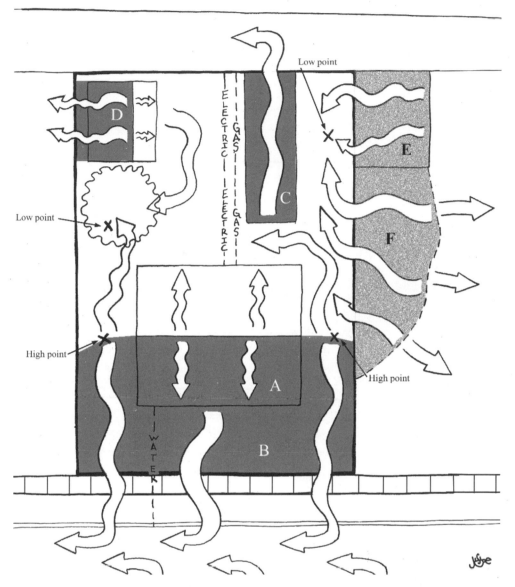

RUNOFF LOST

A. 3,282 gallons runoff lost from 462 sq. ft. half of metal roof. (0.95)

B. 2,246 gallons of runoff lost from 858 sq. ft. section of gravel yard. (0.35)

C. 1,817 gallons of runoff lost from a 270 sq. ft. concrete driveway. (0.90)

D. 538 gallons runoff lost from 80 sq. ft. section of shed's asphalt shingle roof. (0.90)

RUNON GAINED

E. 1,421 gallons runon gained from neighbor's 200 sq. ft. metal shed roof. (0.95)

F. 1,570 gallons runon gained from 350 sq. ft. section of neighbor's compacted dirt yard. (0.60)

Fig. 2.4. Estimated annual runoff volumes off each type of catchment surface are listed with surface material and runoff coefficient. In an average year of 12 inches of rainfall this site receives 32,912 gallons (124,407 liters) of rainfall, gains 2,991 gallons (11,305 liters) of runon from the neighbor's yard and shed roof, and loses 7,883 gallons (29,797 liters) of runoff for a total site rainwater budget of 28,020 gallons (105,915 liters). If the landscape were changed to harvest both the runon and runoff, the site's annual rainwater resources could increase up to a total of 35,903 gallons (135,713 liters). Still more runoff from sidewalk and street could be harvested within the public right-of-way to grow public street trees (see volume 2, the chapter on reducing hardscape, for strategies on harvesting street runoff).

Box 2.5. The Example Site's Water Expenses

Refer to figure 2.4.

The estimated annual water "expense" or needs of the water budget for the Tucson, Arizona site from figure 2.4 is 8,000 gallons per year for the landscape (one mature citrus tree), and 104,248 gallons for the four-person household's interior requirements (washing, bathing, cooking, drinking, toilet, evaporative cooler, with usages based on www.h2ouse.org data for a non-conserving household).

So all the water needs of the current, or a denser, landscape could easily be met by the 35,903 gallons (135,713 liters) of harvestable rainfall and runon, and the landscape's density could be increased still more with the on-site reuse of greywater. Of the household's current interior water needs 6% could be met by harvesting the home's 6,565 gallons of roof runoff. By implementing simple water conservation strategies recommended by h2ouse.org (use of low-flow toilets, faucet aerators, efficient washer, and evaporative cooler without bleed valve) the family could reduce annual interior water needs to 62,684 gallons, with roof runoff providing 10% of that. An additional 7,350 to 20,636 gallons could be conserved with such strategies as planting cooling shade trees to offset cooler use and the use of a composting toilet to eliminate water used for flushing. Lifestyle changes and learning to live with less can lead to still more water conservation. By combining all these conservation strategies, and expanding the roof surface (perhaps with a covered porch) this household could meet all its water needs from rain in an average year. Read on and see what strategies would be appropriate for your site and needs. Every site is unique, and you decide how far you want to take it, using strategies found in this and the next two volumes of this book.

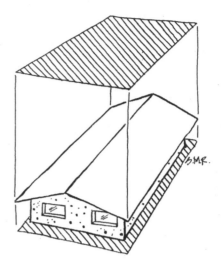

Fig. 2.5. Size of a roof catchment: measuring only the outside dimensions—or "footprint"—the roof's edge. Ignore the roof slope; no more rain falls on a peaked roof than falls on a flat roof with the same footprint.

Note: This book and the rainwater-harvesting principles emphasize the harvest and utilization of localized runoff and runon high in the watershed *before* it enters a drainageway. Once the water is in a drainageway, it is to remain there, although its flow can be slowed to allow for more infiltration as is the case with a check dam.

WHERE TO GET INFORMATION ON RAINFALL RATES AND OTHER CLIMATIC DATA

Seek out data on your area's annual rain and snowfall. Note record high and low temperatures in each season to determine suitable plants to grow. Find out the longest periods of drought and of rainfall to get an idea of the volume of water storage to plan for. Check evaporation rates: The higher the evaporation rate, the more important water-harvesting strategies are that limit evaporation losses. Assess prevailing wind direction and intensity—you may want to plant windbreaks irrigated with harvested rainwater.

Good information sources include:
• In the United States: National Weather Service's website at www.wrh.noaa.gov. Locate the weather

stations closest to your site and find out their elevations. Download data from those stations that are most like your site.

- In Arizona: Arizona Meteorological Network at ag.arizona.edu/azmet. Evaporation rates, prevailing winds, soil temperatures, and minimum/maximum temperatures are listed for various sites. For other states contact your local agricultural extension service for similar meteorological networks.
- The U.S. National Forest Service compiles data for remote weather stations, though the data is not as comprehensive nor standardized as the above two resources. This data can nonetheless be valuable for rural sites since a Forest Service weather station may be closer to a given site than one monitored by other agencies.
- Local airports, since they collect and record climatic data.
- Buy a rain gauge from a hardware or garden store to begin keeping precipitation records for your site.

ESTIMATE YOUR SITE'S WATER NEEDS

Determine the "expense" side of your water budget by estimating your household and landscape water needs. Your water bill reflects current water use. The user-friendly website www.h2ouse.org provides water use rates for household appliances, and recommended conservation strategies. Estimated water needs of plants can be obtained from the local agricultural extension office (or see, for example, appendix 4 "Example Plant Lists and Water Requirement Calculations for Tucson, Arizona"). Better yet, take a hike to observe native plants that grow naturally in *your* area on rainfall alone. Water needs of plants will vary widely depending upon the plant, its size, and the soils, climate, and microclimate in which it is planted. Keeping that in mind, in Tucson, Arizona a mature 20-foot (6-m) tall and wide native mesquite tree will use about 3,000 gallons (11,355 liters) of water per year, while a mature 16-foot (4.8-m) tall and wide exotic non-native citrus tree will use about 8,000 gallons (30,282 liters) per year.[4]

Compare your site's water needs to the volume of rain falling directly on, or flowing through, your site.

How much of your domestic water needs could you meet by harvesting rooftop runoff in one or more tanks? How much vegetation could you support by harvesting rainfall directly in your soil? How can you balance your water budget using harvested rainwater as your primary water source?

ESTIMATE YOUR SECONDARY ON-SITE WATER SOURCE—GREYWATER

Greywater is the water that drains from your household sinks, tubs, showers, and washing machine. It does *not* include the water draining from your toilet, which is called *blackwater*. Ideally greywater's original source is harvested rainwater, but more often it is municipal or well water drawn from the tap. Whatever the source, you can turn this household wastewater into a resource by using it to safely and productively irrigate your landscape. The volume of available household greywater depends on how much water goes down your drains. Every household is different, but the information in box 2.6 offers ballpark estimates of typical volumes used. Add up your site's estimated greywater volume from each source (see figure 2.6 for an example). You'll then use this information in chapter 4 to create an integrated conceptual rainwater-harvesting and greywater-harvesting plan for your site. In volume 2 there's a chapter which provides more specific information on greywater-harvesting systems.

After you have a good idea of your site's rainwater and greywater resources you are now ready to consider some of the strategies presented in chapter 3 with which you can harvest that water. Chapter 4 shows you how to increase the value of these strategies exponentially by integrating them with additional resources including the sun, vegetation, and more.

REAL LIFE EXAMPLE

LIVING WITHIN A DRYLAND HOME'S RAINWATER BUDGET

Matthew Nelson and Mary Sarvak come very close to living within their rainwater budget. You can see much of their annual rainwater supply in the

GREYWATER PRODUCTION

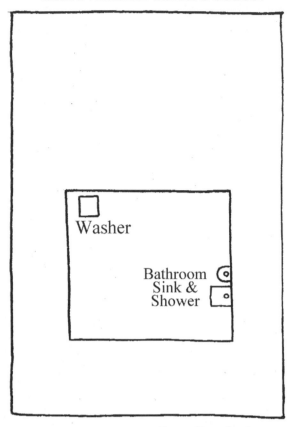

Conventional Washer	Low-flow bathroom Sink & Shower
weekly: 180 gallons 684 liters	
monthly: 720 gallons 1,368 liters	weekly: 302 gallons 1,144 liters
yearly: 9,360 gallons 35,568 liters	monthly: 1,208 gallons 5,776 liters
	yearly: 15,704 gallons 59,672 liters

Fig. 2.6. Example of a home's average estimated greywater resources produced through four residents' use of water. This greywater could be accessed and recycled within the landscape with the installation of a greywater system.

middle of their living room. A 2,500-gallon (9,500-liter) ferrocement cistern rises three feet from a hole in the floor (fig. 2.7). It looks like a *tinaja*—a desert water hole carved into bedrock. Peering into the topless cistern you come face to face with the rainwater that has drained in from the downspout in the middle of their 1,500 square foot (135 m²) roof. That rainwater supply meets all their domestic needs except drinking water. As they use their water, they see its level

Fig. 2.7. Matt and Mary beside their indoor cistern

drop. "You get more conservative as you see the water level fall," says Matt. Though when the water level is high in the middle of the rainy season, they feel free to indulge a little. Talking about their harvested rainwater Mary says, "During the intense heat of July and August, we can walk in the door of our home and splash cool water all over ourselves. Or we can just jump in!"

The system was designed to supply the average water needs of a family of three for four months between rains typical for the site. Their rainfall averages 14 inches a year. A 10-gallon RV pump pressurizes the water and sends it throughout the house to all the sinks, the bathroom, and the washing machine.

Outside, 100% of the landscape's water needs are met by rainwater. When the home was built, extreme care was taken to avoid disturbing or destroying any of the existing native vegetation beyond the footprint of the house. As a result, nothing was spent on landscaping—it already existed. In addition, 98% of the landscape has never needed supplemental irrigation because it was already well established and perfectly adapted to natural rainfall. The 2% of the landscape needing irrigation includes an apple tree planted after the home was built and some potted plants. This irrigation is done entirely with recycled rainwater in the form of greywater from the drains of their washing machine and kitchen sink.

Matt and Mary's rainwater oasis in the Sierrita Mountains of southwest Arizona is located in a boondoggle development from the 1970s that sold home

Box 2.6. Estimates of Household Greywater Resources.[5]

Based on figures from "Branched Drain Greywater Systems" by Art Ludwig: www.oasisdesign.net and www.greywater.com

Fixture	Frequency Used	Volume Used	Weekly Use Per Person	Yearly Use Per Person
Top-loading washing machine	1.5 uses/person/wk	30 gal/use 114 liters/use	45 gal/wk/person 171 liters/wk/person	2,340 gal/yr/person 8,892 liters/yr/person
Bathtub	1.5 uses/person/wk	20 gal/use 76 liters/use	30 gal/wk/person 114 liters/wk/person	1,560 gal/yr/person 5,928 liters/yr/person
Shower	5 uses/person/wk	13 gal/min 49 literes/min	65 gal/wk/person 247 liters/wk/person	3,380 gal/yr/person 12,844 liters/yr/person
Bathroom sink	21 uses/person/wk	0.5 gal/min 1.9 liters/min	10.5 gal/wk/person 39 liters/wk/person	546 gal/yr/person 2,074 liters/yr/person
TOTAL			150.5 gal/wk/person 571 liters/wk/person	7,826 gal/yr/person 29,738 liters/yr/person

Box 2.7. Estimates of Low-Water-Use Front-Loading Washing Machine Greywater Resources.

Based on figures from www.greywater.com. Compare to the data for a top-loading washer in box 2.6.

Fixture	Frequency Used	Volume Used	Weekly Use Per Person	Yearly Use Per Person
Front-loading washing machine	1.5 uses/person/wk	10 gal/use 38 liters/use	15 gal/wk/person 57 liters/wk/person	780 gal/yr/person 2,964 liters/year/person

sitcs to folks sight-unseen. No utilities were installed and few sites were ever developed. At sites that were eventually developed, neighbors run generators around the clock for electricity, and nearly all truck in their water, ignoring rain that falls freely from the sky. Matt and Mary generate most of their electricity from solar panels on their roof, and aside from drinking water they have had to haul in water only once, during a five-month drought.

Designing around the assessed on-site resources of rain and sun is what has enabled Matt and Mary to live with modern comforts while consuming and paying for just a fraction of the off-site resources their neighbors consume. Matt and Mary are also directly connected with their beautiful desert surroundings,

the source of their resources, and their need to keep consumption in balance. These are the ideas I encourage others to mimic, and why I feature Matt and Mary here. However, two of the specific techniques used on their site should be altered, rather than duplicated. First, a closed exterior tank would be superior to the open living room tank. That is because, while the open water beneficially cools the home in summer, it also cools it in winter as water pours in from frigid downpours (window and door screens along with ten mosquito-eating fish take care of the potential mosquito problem associated with thc opcn watcr). Second, Matt and Mary are concerned that the elastomeric paint on their roof may taint their water. This is the reason they haul drinking water in from town.

I recently told them about various elastomeric paints and roof coatings approved for use with potable rainwater catchment systems (see the chapter on cistern components in volume 3). With the application of one of these coatings (and cooling their bodies in the rainwater-fed shower, rather than plunging into the cistern), they'll be worry-free and rainwater-rich (but winters will still be cold).

Strive to live in balance with your site as Matt and Mary do by first assessing your resources and needs— the first principle of water harvesting: *Long and thoughtful observation*. And as you plan what strategies and techniques you use to harvest the site's resources persistently think of how the system will work as a whole—the last principle of water harvesting: *Continually reassess your system: the "feedback loop."* The better your assessment the better your system will be, and the following chapters give you more tools to do just that.

CHAPTER

Overview: Harvesting Water with Earthworks, Tanks, or Both

Once you've estimated your on-site water resources and needs (from the previous chapter), the next step is to answer the following question: How do you plan to use your water resources? How you answer that question will largely determine the best strategy to harvest that water—be it water-harvesting earthworks, tanks, or both. This chapter compares these strategies, and briefly describes some of their more detailed techniques and applications to help you visualize and decide which are most appropriate for your unique site. (*These strategies are covered in detail in volumes 2 and 3.*) Basic recommendations are then given for harvesting water within a residential landscape. The chapter wraps up with a homesteading couple harvesting water in earthworks and tanks. The next chapter (Integrated Design) then shows you how to integrate these strategies with other aspects of your site so your water-harvesting efforts can capitalize on additional free resources such as the sun's light, power, and winter heat; provide privacy, wildlife habitat, summer shelter, shade, and cooling; filter out noise, light, and air pollution; while also controlling flooding and erosion—in other words, how to maximize your site's potential!

HOW DO YOU PLAN TO USE YOUR HARVESTED WATER?

LANDSCAPE OR GARDEN USE

If you plan to use your harvested water for landscape or garden use, begin harvesting water in the soil using *earthworks*. Landscapes that harvest water in the soil and are planted with low-water-use native plants can often subsist on rainfall alone without supplemental irrigation from a tank or tap, once the vegetation is established. However, tanks give you the option of applying supplementary irrigation in dry times—especially if a vegetable garden or less hardy non-native vegetation is planted.

POTABLE USE AND WASHING

If you plan to use your harvested water for potable use and washing, begin harvesting water in *tanks*. But do not forget the soil, and continue to direct overflow from the tank, greywater from your house, and the runoff from the general landscape into water-harvesting earthworks. While a cistern can be installed to irrigate a garden or supplement a landscape in dry times, the more water you can effectively harvest and hold in the soil, the less supplemental cistern irrigation will be needed.

MATCH THE QUALITY OF WATER BEING HARVESTED TO THE STRATEGY USED TO HARVEST IT

Your site's highest quality rainwater, typically runoff from clean roof materials such as metal, slate, tile, or elastomeric paints approved for rainwater collection systems, is the most appropriate for storage within cisterns, domestic consumption, or use on vegetable gardens. Stormwater from dirtier surfaces such as earthen slopes, streets, or sidewalks should be directed to trees and shrubs within passive water-harvesting earthworks. Household greywater should be directed to and utilized within mulched basins planted with trees and shrubs. Do not store greywater in tanks.

Keep in mind the question "How do you plan to use your harvested water?" as you read the comparisons of these water-harvesting approaches in box 3.1.

RECOMMENDATIONS FOR HARVESTING WATER FOR A LANDSCAPE AROUND A HOME

I recommend that all my clients *begin* with water-harvesting *earthworks*. I stress the importance of placing earthworks and the plants they support in areas where they will provide beneficial functions, such as the east and west sides of a home for passive cooling with shade trees. I stress using a low-water-use native plant palette of indigenous vegetation found within a 25-mile (40-km) radius of the site and 500 feet (150 m) above or below the site's elevation. Some sites may require defining *native* with a larger radius to bring in more diversity, but start with the small radius to ensure you do not overlook superior local species. Such plants are typically the most beneficial for native wildlife, and create a Sense of Place rooted to our local bioregion. These plants are adapted to local growing conditions and can be cut off from supplemental irrigation once the vegetation is well established, or has grown to a desired height. Such native plant landscapes are beautiful and low maintenance.

I recommend *cisterns to complement* the foundation of water-harvesting earthworks laid throughout the landscape or garden. The cisterns stretch the availability of the rain long into the dry season—especially for more water-needy landscapes and vegetable gardens. The earthworks utilize the overflow water from the tanks, and make all the water used within the landscape or garden go further.

Fruit trees or other water-needy vegetation, if used, should be placed close to the house to create an oasis-effect around the home. The plants are then supported by the rainwater from the roof; greywater from the sinks, showers, tubs, and washing machine; and the care of those living inside the house. *The goal is to select and plant the vegetation in such a way that rainwater will be the landscape's primary source of water, greywater its secondary source, and imported water will only be used as a supplementary source in times of need.* Sprawling water-needy vegetation placed far from the house makes achieving this goal far more difficult. The plants will be out of sight and out of mind, neglected and underutilized by you, and potentially too distant for easy irrigation from roof runoff or household greywater. Instead, hardier native vegetation goes to the periphery of the property, and gets planted within water-harvesting earthworks. This hardy vegetation can shelter the less hardy plants, such as the fruit trees, from excessive sun or wind. Hardy native plantings placed along the periphery support more native wildlife, which appreciates less interference from us. (See figure 3.1.) And of course, a low-water-use native landscape placed in water-harvesting earthworks is easy and inexpensive to establish and maintain. (See appendix 4 for an example plant list for Tucson, Arizona, and also volume 2, the chapter on vegetation, for more planting tips.)

SAMPLING OF STRATEGIES/TECHNIQUES

Now that you have some idea of how you want to harvest your rainwater—in the soil, tanks, or both, you are ready for specifics! In volume 2 of *Rainwater Harvesting for Drylands*, the chapters describe in detail how to harvest rainwater in the soil using various earthwork techniques. The chapters in volume 3 address harvesting rainwater in tanks or cisterns. Below, I give a brief overview/sampling of some of the techniques. The illustrations in this chapter are for

Box 3.1. Comparing Earthworks and Tanks

CHARACTERISTICS	EARTHWORKS	TANKS
Water uses	Provides large quantities of high quality rainwater to garden and landscape	Provides water for drinking, washing, fire control, and supplemental irrigation. Water quality will vary with catchment surface, tank construction, screening, maintenance, and first flush system. Rainwater has very low hardness.
Water collection areas	Can collect water from roofs, streets, vegetation, bare dirt, greywater drains, air conditioner condensate, etc.	Need a relatively clean collection surface (typically a metal, tiled, or slate roof) located higher than the tank
Water storage capacity	Very large potential to store water in the soil	Storage capacity limited by the size of the tank
Cost	Inexpensive to construct and maintain. Can build with hand tools, though earthmoving equipment can speed up the process	Much more expensive than earthworks to construct and maintain. Cost varies with size, construction material, above- or below-ground placement, self-built or prefab, etc.
Location	Do not locate within 10 feet of wall or building foundation. May be difficult to use in very small yards with adjacent large roofs	Can locate within 10 feet of wall or building foundation, but you must be able to walk around entire above-ground tank to check for, and repair, leaks. Tanks increase water-storage potential in very small yards.
Time period water is available	Water available for limited periods after rainfall depending on soil type, mulch, climate, and plant uptake	Water is available for extended periods after rainfall.
Maintenance	Earthworks work passively; require some maintenance after large rainfalls	Maintenance required; must turn valve to access water and may need pump to deliver water
Erosion control	Very effective for erosion control	Can assist with erosion control
Greywater collection	Very effective at harvesting greywater from household drains	Not appropriate to harvest greywater in tanks due to water-quality issues. Never store greywater in a rainwater tank.
Water quality impacts to environment	Pollutants in greywater and street runoff intercepted in the soil stay out of regional waterways	Less impact than earthworks to the broad environment
Impacts on urban infrastructure and flooding	Can capture large volumes of water, reducing need for municipal water, stormwater drains, stormwater treatment, and decreasing flooding	Can capture low to moderate volumes of water, reducing demand for municipal water, stormwater drains, stormwater treatment, and decreasing flooding
Groundwater recharge	Can sometimes recharge shallow groundwater tables	Not an efficient use of tank water. However, use of cistern water instead of municipal/well water reduces groundwatrer depletion.

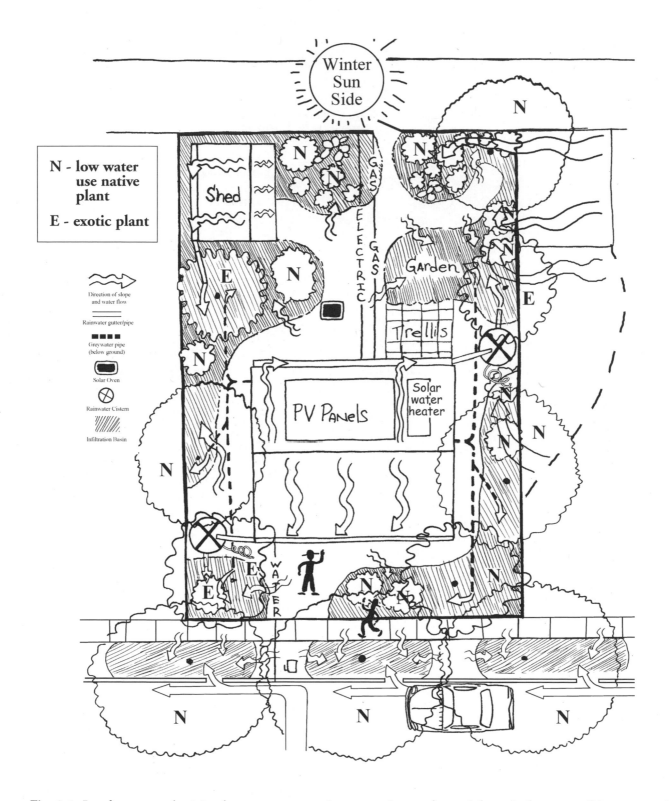

Fig. 3.1. Landscape emphasizing low-water-use native vegetation at the periphery, in less accessible areas, and in areas of less water. Needier exotics are placed close to the home where roof runoff, household greywater, and attention are easy to access. However, a lower maintenance landscape could consist entirely of low-water-use natives.

you, the reader, to conceptualize earthworks and water storage as you formulate an integrated water-harvesting plan for your site—and to whet your appetite for more in volumes 2 and 3.

OVERVIEW OF WATER-HARVESTING EARTHWORKS

The techniques outlined below are discussed in detail in volume 2. However, appendix 1 in this volume has illustrations of the use of many of these techniques in controlling various erosion patterns.

Berm 'n basin

A berm 'n basin is a water-harvesting earthwork laid perpendicular to land slope, designed to intercept rainwater running down the slope and infiltrate this water in a localized area, usually the soil and root zone of existing or planned vegetation. It usually consists of two parts: an excavated basin and a raised berm located just downslope of the basin. The berm can be made of earth excavated to form the basin or it can be made from brush, rock, or additional earth. The basin holds water; adding the berm enables you to harvest even more water. Use this strategy on sloped land up to a 3:1, 18 degree, or 32.5% grade. Size for the maximum stormwater event. This strategy is not appropriate in drainage ways. (See figures 3.2, 3.3, and 3.4.)

Terrace

A terrace, sometimes called a *bench*, is a relatively flat "shelf" of soil built parallel to the contour of a slope. Unlike berm 'n basins, terraces do not have depressions built into them, so in order to retain rainwater and accumulate organic material, they need to be bordered by berms or low walls. Terraces can be used for gardens, orchards, and other plantings in drylands. Well-built terraces also help control erosion on slopes. Use on sloping land up to a 2:1, 26 degree, or 48.8% grade. Terraces on slopes exceeding a 3:1 grade likely need a retaining wall for stabilization. Size for the maximum storm event. Terraces are not appropriate in soils prone to waterlogging where rainwater infiltration could lead to saturated subsurface

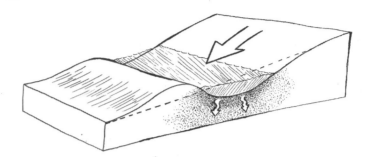

Fig. 3.2. A berm 'n basin holding runoff water and infiltrating it into the soil

Fig. 3.3. A contour berm doubling as a raised path

Fig. 3.4. A series of boomerang and contour berms on a sloping landscape. Arrows denote water flow spread and infiltrated by the earthworks.

Fig. 3.5. Terraces stabilized with salvaged concrete retaining walls in the sloped backyard at the Zemach residence, Los Alamos, New Mexico

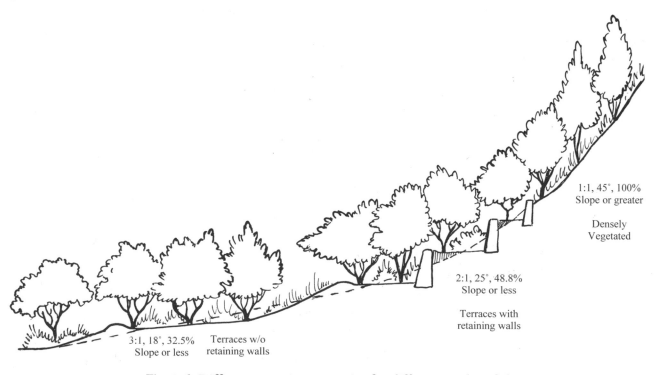

1:1, 45°, 100%
Slope or greater

Densely
Vegetated

2:1, 25°, 48.8%
Slope or less

Terraces with
retaining walls

3:1, 18°, 32.5%
Slope or less

Terraces w/o
retaining walls

Fig. 3.6. Different terracing strategies for different grades of slope

conditions due to the presence of layers—such as clay—that impede the movement of water further down through the soil. (See figures 3.5 and 3.6.)

French drain

A French drain intercepts rainwater into a trench or basin filled with porous materials including gravel, pumice stone, or rough organic matter. These materials have ample air spaces between them that allow water to infiltrate quickly into the drain and percolate into the root zone of the surrounding soil, while creating a stable surface you can walk on. A French drain can be constructed with or without perforated pipe placed horizontally within the porous material that fills the drain. Use on flat to gently sloped land; drains are appropriate where deep subsurface irrigation of landscapes is needed. It is important that the French drain only harvest runoff water that is relatively sediment free, such as from a roof gutter downspout, or edge of paved patio in order to prevent premature silting up of the structure. French drains are not appropriate in drainage ways, areas with sediment-laden water, or beneath or across roadways. (See figures 3.7 and 3.8.)

Infiltration basin

An infiltration basin is a landscaped level-bottomed, relatively shallow depression dug into the earth that intercepts and infiltrates rainfall, runoff, and greywater in the planting basin it creates. This technique works best on flat landscapes where it will have no berm, so all surrounding runoff can drain into it. It can also work on moderate slopes as a terraced basin, water-harvesting tree wells, or interconnected stormwater basins. Size for a maximum storm event and peak surge of greywater. Infiltration basins are not appropriate in drainage ways, areas of shallow groundwater where they might result in standing water, or over septic drain fields. (See figures 3.9 through 3.12.)

Fig. 3.7. French drain infiltrating intercepted runoff from a roof and patio

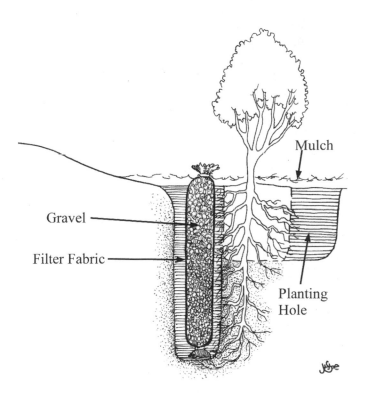

Fig. 3.8. Rock tube variation of a French drain, in which angular rock wrapped in a tube of porous landscape fabric directs water deeper into the soil, encouraging deeper, more drought-hardy root growth

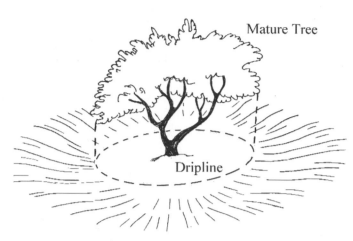

Fig. 3.9. A series of infiltration basins intercepting and infiltrating rainfall and runoff from adjoining street and footpath. Designed right, these basins can act as the sole irrigation system for the associated vegetation, while doubling as a flood control system.

Fig. 3.10. Ideally, the infiltration basin diameter is at least 1.5 times (and up to 3 times) the diameter of the associated *mature* plant's canopy dripline, since roots spread, and *most* water used by plants is drawn from the root zone *outside* the canopy drip line.[1] However, even an undersized basin is better than no basin.

Fig. 3.11A. Rainfall collected in newly constructed infiltration basins minutes after a large summer storm. The basins have not yet been mulched or planted. Milagro Cohousing, Tucson, Arizona
Credit: Natalie Hill

Fig. 3.11B. Same basins mulched and vegetated. Basins are designed to infiltrate water quickly so there are no problems with mosquitoes or anaerobic soils. These basins, with their spongy mulch and soil-burrowing plant roots, infiltrate all water within 20 minutes.

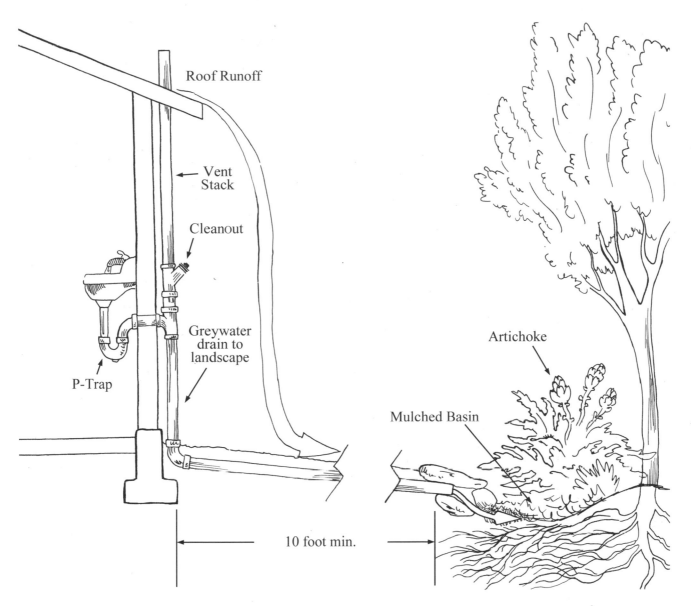

Fig. 3.12. Roof runoff and bathroom sink greywater directed to a well-mulched and vegetated infiltration basin. Although not shown, the basin continues around to the other side of tree. Note P-trap and vent stack between interior drain and exterior greywater outlet, which prevents potential odor and insect entry into house.

In the figure, the following labels appear: Roof Runoff, Vent Stack, Cleanout, Greywater drain to landscape, P-Trap, 10 foot min., Artichoke, Mulched Basin.

Imprinting

Imprinting is a water-harvesting technique used to accelerate the revegetation of disturbed or denuded land with annual precipitation from 3 to 14 inches (76 mm to 330 mm) by creating numerous small, well-formed depressions in the soil that collect seed, rainwater, sediment, and plant litter, and provide sheltered microclimates for germinating seed and establishing seedlings. Ideally, each imprint captures enough water to germinate one or more seeds and sustain their growth. Imprints have enough water-storage capacity to increase infiltration to levels above most dryland rainfall rates, thus eliminating nearly all runoff and associated erosion. The imprints are V-shaped and are made with an imprinter roller, though hoof prints, slow-moving deep-knobbed wheels, or bulldozer tracks can create somewhat similar effects. Sites should be at least an acre to justify

Fig. 3.13A. An imprinter roller pulled by a tractor creating and seeding imprints on barren, compacted earth, Marana, Arizona

Fig. 3.13B. Native vegetation restoring the land and soil, post imprinting

bringing in equipment. The slope may be up to 2:1, 26 degree, or 48.8% grade. Individual imprints are about 4 to 7 inches (10 to 18 cm) deep, 10 inches (25 cm) long, and 8 inches (20 cm) wide. Imprinting is not appropriate in drainage ways. (See figure 3.13.)

Mulching

Mulching is the application of porous materials such as compost, aged manure, straw, or wood chips onto the surface of the soil. Mulch is both a spongy welcome mat luring water into the soil and a shelter-ing cover reducing soil-moisture loss to evaporation. It also limits soil erosion and weed growth, while improving soil fertility.

Mulch is not mixed or dug into the soil, since most mulch materials are rich in carbon and could deplete nitrogen in the soil as they decompose. Nitrogen depletion occurs because microorganisms assisting decomposition need nitrogen to break down carbon, which they get from the surrounding soil. Carbon-rich mulch applied on the soil's surface decomposes more slowly than if it's mixed into the soil, so soil nitrogen is consumed at a slower rate that does not adversely affect vegetation.

Increase the water-harvesting potential of other water-harvesting earthworks by mulching the soil's surface around vegetation and in basins. Mulch also can delineate planting basins from paths. It works best in flat and gently sloped areas, though it can be placed on steeper slopes when combined with other earth-works. Mulch is not appropriate in drainageways. It is important to slow or stop runoff before it comes into contact with mulch on slopes, and to keep mulch close to buildings. (See figures 3.14 and 3.15.)

Reducing hardscape and creating permeable paving

Hardscape reduction is a strategy to minimize the need for pavement through creative planning and design, and the removal of impervious pavement, where possible. By reducing impervious hardscape (and sloping it toward adjoining earthworks), you can increase the adjoining pervious areas to enhance on-site water infiltration of the hardscape's runoff, reduce runoff leaving the site that would otherwise contribute to downstream flooding and contamination, and decrease the heat-island effect caused by excessive, exposed hardscape.

Permeable paving is a broad term for water-harvesting techniques that use porous paving materials to enable water to pass through the pavement and infil-trate into soil, passively irrigating adjoining plantings, dissipating the heat of the sun, reducing soil com-paction, allowing tree roots beneath the paving to breathe, filtering pollutants, and decreasing the need for expensive drainage infrastructure. As a water-

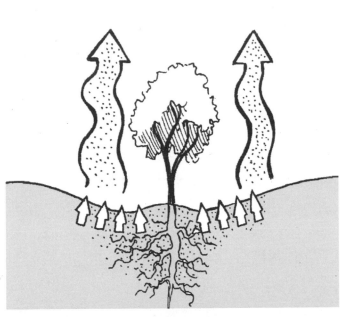

Fig. 3.14A. A basin without mulch losing the bulk of the soil moisture to capillary action and evaporation. Note the small stressed tree.

Fig. 3.14B. The basin mulched for improved infiltration and retention of water into the soil

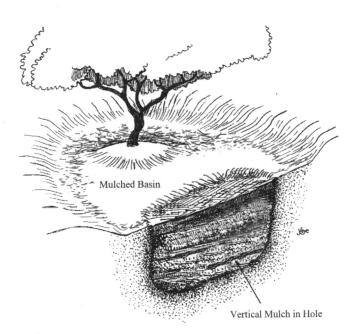

Mulched Basin

Vertical Mulch in Hole

Fig. 3.15. Vertical mulch variation (mulch-filled hole or trench) encouraging infiltration and retention of water deeper into the root zone of the soil

harvesting technique, it is most useful on densely developed sites with little unpaved surface. Permeable paving is most effective when it only harvests the rainwater that falls directly upon it, without additional runoff from upslope areas. It is not appropriate in drainage ways. Raise permeable paving above the surrounding landscape to prevent settling or pavement displacement due to poor draining subsoil, and prevent sediment-laden stormwater runoff from plugging pores in permeable pavement. (See figures 3.16, 3.17, and 3.18.)

Diversion swale

A diversion swale is a gradually sloping drainageway that slowly moves water from one point to another. Like a berm 'n basin, it usually consists of a generally linear basin with the excavated earth placed downslope to form a berm. Unlike a berm 'n basin, which is constructed on-contour to contain and allow water to soak into the earth locally, a diversion swale is built slightly off-contour, allowing a portion of the water to soak into the soil locally while moving surplus water slowly downhill from one place to another, infiltrating water all along the way.

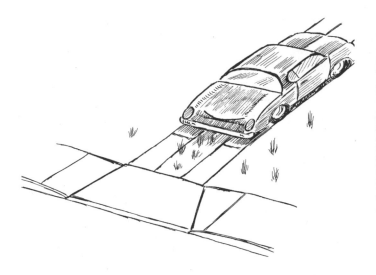

Fig. 3.16. A two-track driveway can reduce impermeable surface area by 60% compared to conventional concrete driveways.

Fig. 3.17. Narrower streets and young native food-producing mesquite trees irrigated by harvested street runoff, Civano, Tucson, Arizona

Diversion swales are used to intercept, infiltrate, and redirect both sheet flow and channelized water. Diversion swales can tame the force of water that rushes out from a culvert or roadside bar ditch, transforming the concentrated fast-moving water into a valuable resource by spreading out and calming the flow. Diversion swales can direct runoff to a water-harvesting berm 'n basin, infiltration basin, pond, or other final destination.

Diversion swales are not appropriate in drainage-ways, and they should not be used in alkaline soils prone to salt buildup and waterlogging. (See figures 3.19 through 3.22.)

Check dam

A water-harvesting check dam is a low barrier that is *permeable* or "leaky," and placed perpendicular to the flow of water within a drainage. The check dam does not stop, but rather *slows* the flow of water. As it does so, running water temporarily backs up behind the dam and spreads out over more of the drainage's surface before flowing through and over the dam. By slowing and spreading the flow of water, check dams help moisture infiltrate into the soil, reduce down-stream flooding by moderating the peak flow of water, retain soil and organic matter on the upslope side of

Fig. 3.18. A small yard with hardscape kept permeable by installing recycled sidewalk chunks with ample gaps for water infiltration, Amado residence, Tucson, Arizona

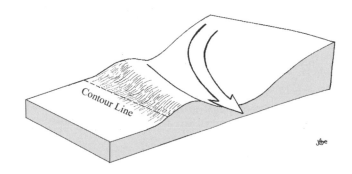

Fig. 3.19. Diversion swale

Fig. 3.20. A series of diversion swales as a water-harvesting overflow route from one infiltration basin to another

Fig. 3.21. Diversion swale speed humps directing runoff from the road to plantings

the dam, reduce erosion, and stabilize a section of the landscape. A check dam is built across a drainage that only flows periodically, and is often used to help heal an eroding arroyo or gully. Check dams can improve roads or paths where they cross ephemeral drainages. Build check dams no taller than 3 feet (0.9 m). Do not place them in or just downstream of a curve in a drainageway, or in the narrowest point or just downstream of a constriction in a drainageway. (See figures 3.23, 3.24, 3.25.)

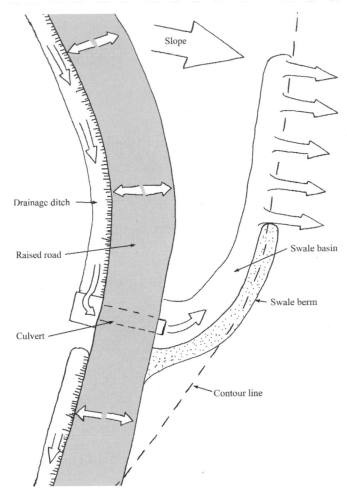

Fig. 3.22. Diversion swale harvesting stormwater and calming its flow once discharged from a culvert crossed by the dirt road

Vegetation

Vegetation is the life that emerges from a water-harvesting system. It increases water infiltration and soil stabilization with root penetration, and provides multiple resources and benefits. Vegetation is encouraged and used within or beside every water-harvesting earthwork. Vegetation's use is appropriate throughout all kinds of watersheds, with no limit to the applicable slope. It is important to plant climate-appropriate species at densities that, once established, can subsist primarily, or exclusively, on the site's harvested rainfall. Locate and space all plantings according to their expected mature size, not their size at the time of planting. (See figures 3.26 and 3.27; there is more on vegetation's use in integrated design in the next chapter.)

Fig. 3.23. A gabion is a wire-wrapped stone check dam. A wire-wrapped downstream apron prevents erosion at the base of the check dam and dam failure.

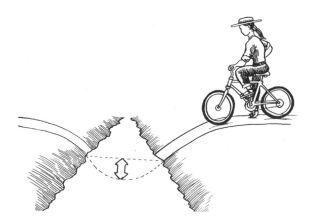

Fig. 3.24A. Deep impassable flow without a check dam

Fig. 3.24B. Spread out, shallow, and passable flow with check dam

Fig. 3.27. Rain-irrigated cactus fruit and pads. Cactus pads can be cooked as a vegetable.

Fig. 3.25. A series of small check dams with low point in the middle of the channel harvest water and check erosion in a roadside drainage ditch. The road is to the right. Credit: Ann Phillips

Higher Water Needs /Tolerance

Lower Water Needs /Tolerance

Lowest Water Needs /Tolerance

Fig. 3.26. Planting according to water needs and tolerance

OVERVIEW OF CISTERN SYSTEMS

Cistern systems should follow the rainwater-harvesting principles described in chapter 1, but because they store a readily accessible body of water, you should also follow an additional set of principles specific to cisterns. The principles and components of cistern systems outlined below are discussed in detail in volume 3, as are various tank options (premanufactured or made on site). Go online to watertanks.com for a quick reference to some available pre-manufactured tanks and current costs.

TEN BASIC COMPONENTS OF A SIMPLE ABOVE-GROUND RESIDENTIAL CISTERN SYSTEM (FIG. 3.28)

1. Catchment surface
2. Gutters and downspouts
3. Screening of cistern and downspout openings
4. First-flush systems (optional)
5. Cistern
6. Vent
7. Overflow
8. Faucet and valve
9. Filters and pumps (optional and not pictured)
10. The maintenance team: You

NINE CISTERN SYSTEM PRINCIPLES

See figures 3.30 through 3.36 for examples of cisterns/tanks. See volume 3 of *Rainwater Harvesting for Drylands* for more details on how to meet the cistern principles.

1. **Ensure adequate inflow.** Don't lose water. Size your gutters, downspout, and inflow pipe to handle the maximum rainfall intensity likely to occur in your area. (See the downspout sizing information in volume 3, the chapter on cistern components.)

2. **Ensure adequate outflow and use it as a resource.** The diameter of the cistern overflow pipe must be equal to or larger than the diameter of the cistern's inflow pipe so your system does not back up. Direct that overflow resource to another tank or mulched and vegetated infiltration basins.

3. **Design your system to collect high quality water.** The higher the quality of harvested water the more options you have for its potential use, so don't contaminate your water with any toxic or contaminated materials making up your system. Materials rated for contact with potable water yield the highest quality water.

4. **Design a closed system that passively filters itself.** Design or install "closed" cisterns screened off from sunlight, insects, and critters so algae and bacteria will not grow, mosquitoes will not propagate, and drowned critters or their waste will not contaminate your water. Additionally, tank covers will reduce water loss to evaporation. Construct the outflow pipe from the cistern (the "supply" pipe) a minimum of 4 inches (10 cm) above the bottom of the cistern to keep the sludge of sediments (leaf litter, dust, etc.) from being pulled into the supply pipe. See figure 3.29 for an example.

5. **Maintain access to your tank and its interior.** You need access to check water levels, clean out the tank, and make repairs. Place above-ground cisterns so there is enough space to walk completely around them to check for (and repair) leaks and conduct inspections, especially if they are close to a building.

6. **Vent your tank.** All covered tanks with tight-fitting lids or tops must be vented to prevent a vacuum from forming within the tank when large quantities of water are quickly drawn from the tank.

7. **Use gravity to your advantage.** Place your tank at a location where you can utilize the elevation of the catchment surface and the free power of gravity to collect rainwater and distribute it around your site. Below-ground tanks may not be able to use gravity to distribute the stored water, but they must be designed so overflow can occur with gravity. You can always add a pump to increase water pressure and performance if needed, but don't turn a pump into a crutch your system must depend on.

continues on page 75

Fig. 3.28. Basic components of a cistern system

KEY:

1. Catchment surface
2. Gutters and downspouts
3. Screening of cistern and downspout openings
4. First-flush systems (optional)
5. Cistern
6. Vent
7. Overflow
8. Faucet and valve
9. Filters and pumps (optional and not pictured)
10. The maintenance team: You

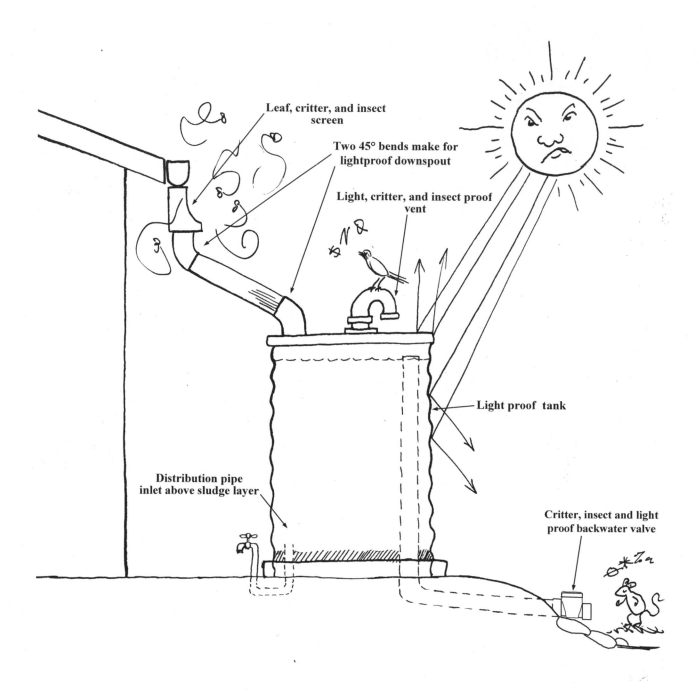

Leaf, critter, and insect screen

Two 45° bends make for lightproof downspout

Light, critter, and insect proof vent

Light proof tank

Distribution pipe inlet above sludge layer

Critter, insect and light proof backwater valve

Fig. 3.29. A closed cistern system

Fig. 3.30. Galvanized tank and wet system downspout, Tank Town, Texas

Fig. 3.32. Culvert tanks with light-, insect-, and critter-proof lids. Overflow moves from one tank to the next, then out to vegetated basins not shown. Milagro Cohousing, Tucson, Arizona

Fig. 3.31. A 10,000 gallon (38,000 liter) steel tank/mural and dry system downspout collecting roof runoff for irrigation, Children's Museum, Santa Fe, New Mexico

Fig. 3.34. A 6,000 gallon (22,680 liter) ferrocement tank storing roof runoff for domestic use and irrigation, Meuli residence, Tijeras, New Mexico. The square piece of corrugated metal atop the tank is the lid to the access hole.

Fig. 3.33. A 305 gallon (1,159 liter) polyethylene tank with rainhead downspout screen collecting roof runoff from porch roof for courtyard landscape irrigation, Nature Conservancy offices, Tucson, Arizona. The two 45° bends in downspout pipe keep direct light out of tank. Access hole lid includes subtle vent. At bottom of tank on left side, a ball valve accesses tank water; beside the valve is a spigot accessing municipal water. Beside the spigot is the backwater valve directing overflow water to a mulched basin.

Fig. 3.35. A 7,000 gallon (26,460 liter) poured-in-place concrete tank faced with stone doubles as a patio and collects roof runoff for irrigation of citrus trees and other plants below. Tucson, Arizona.

8. **Make rainwater use convenient**. Where feasible, select your tank location so it is near both the water source (roof) and the destination (garden, sink, etc.). This will minimize the length of downspouts, pipes, and hoses, which will save money and materials and help maintain water pressure. At the very least, place or plumb the cistern's faucet conveniently close to your point of use, even if the tank is inconveniently distant.

9. **Select and place your cistern so it does more than store water**. The more your cistern does, the more cost-effective it is. By designing a cistern to also act as part of a privacy screen, fence, retaining wall, or support pillar for a covered porch you eliminate the cost of buying other materials to make that section of screen, fence, wall, and/or pillar.

REAL LIFE EXAMPLE

THE RUNNING RAIN SOCIETY

I end this chapter with the story of a homesteading couple who harvest rainwater in both soil and tanks. Although they live on 40 acres (16 ha) of land, the principles they learned from living many years with an evolving system can apply equally well to a 1/10th-acre urban lot, an entire community, and everywhere in-between.

LIVING OFF RAINWATER HARVESTED IN TANKS AND SOIL

Twenty-seven years ago, Dan Howell found his ticket from the Southern California rat race to the country life. There on the laundromat bulletin board a 3 x 5"card read, "40 acres in beautiful, rural New Mexico—cheap."

Dan bought it sight unseen and set out to find his land in west-central New Mexico. At 7,000 feet elevation (2,133 m) the air was crisp beneath an expansive blue sky. The land was starkly beautiful but eroding. Hardy pinyon pine, juniper, and thin grasses sparsely peppered the land speaking of climate extremes. Summer day temperatures would soar over 100° F (38° C), and winter nights would drop well below

freezing. Above all else, it was dry. Annual rainfall averaged just 14 inches (355 mm). Ground-water was so deep almost everyone in the area brought their water in on trucks. Soils were denuded from poorly managed cattle grazing and excessive timber cutting. Without the anchoring effect of vegetation, wind and stormwater ran over the land, taking the topsoil with them. Dan was determined to move to this land and live in a better way. He went back to Los Angeles to prepare. Five years later, in 1977, he and his new wife Karen headed to the land with a nest egg of money and determination.

Dan and Karen parked their trailer and built a small shed covered with 200 square feet of corrugated metal roofing. A gutter was attached, and drained rainwater to ten sealed 55-gallon (209-liter) plastic drums. When the July rains came, the water barrels filled instantly, allowing Dan and Karen a welcome break from hauling water. Their neighbors hadn't collected a drop of the rain and were back on the long road to town hauling water the next day.

Dan and Karen knew rain and snow would be erratic, ranging from 7 to 30 inches (177 to 762 mm) of precipitation a year, with an average annual rainfall of 14 inches. Over their 40 acres (16.16 ha) of land this range of rainfall would yield 7,623,000 to 32,670,000 gallons (34,654,158 to 148,517,820 liters) of rainfall per year!

The Howells realized that rainfall was far more abundant than the volume of local municipal water they could haul. It was also free—and came to them! So the Howells decided to make rainwater the main water source for all their water needs, and named their homestead the "Running Rain Society."

Once they hooked up a 500-gallon galvanized steel tank to the gutter of their newly built home's metal roof, the water truck got a rest. That one tank, and the new 400-square-foot roof, were enough to provide all the Howells' domestic water needs for drinking, washing, bathing, and cooking for an entire year! (See figure 3.36.)

Fig. 3.36. The Howells' home and 500-gallon
galvanized steel cistern

LIVING ON RAINWATER: TANK STORAGE

The Howells have been living on rainwater for over 20 years. It tastes great and has never needed to be filtered, though they did install a drip irrigation "Y-filter screen" three years ago for the sake of their guests, because tiny rust particles from the tank had started showing up in the water. There is very little air pollution in their area, and more importantly Dan and Karen's metal roof and whole water catchment system is kept clean and toxin-free from start to finish. They have been vigilant in avoiding potential sources of toxins such as asphalt roofing shingles or lead flashing. And they ensure that their tank is closed off to any sunlight, insects, or critters that could introduce or breed unwanted bacteria or diseases.

The Howells have made dramatic lifestyle changes tied to their water use. Dan and Karen use only 5 gallons of water each per day within the home. Any water going down the sink or bath, drains outside to water plants. They built a waterless composting toilet to eliminate water consumption for sewage treatment, and use the composted humanure to enhance their land's fertility. By living within the limits of their on-site water resources the Howells don't impoverish their area from over-consumption, nor do they need to work full time to meet their basic needs.

As a guest cottage and workshop were built, the roof catchment surface grew from 400 to 2,000 square feet (36–180 m²). Two additional tanks raised the Running Rain Society's domestic water storage capacity to 4,500 gallons (17,100 liters). Dan and Karen

started small and worked the kinks out of their system before expanding, and that's just how they recommend others proceed.

HOW THE HOWELLS DEAL WITH FROZEN WATER IN THEIR CISTERNS

In winter, the Howells drain enough water from their 500-gallon (1,900-liter) cisterns so the top 20% of the tanks are empty. This ensures that there is enough room for the water to expand if it freezes. One to two days is the longest Dan and Karen have gone without access to water that was frozen inside their outdoor tanks. They always have a back-up water supply stored inside the home for such occasions. That is the extent of the Howell's preparation for winter water storage, and it works well for them.

(Under colder conditions different strategies may be needed. To prepare for freezing weather, ask water-harvesting locals what they do, and look into local codes for water tank and plumbing installation in your area.)

HARVESTING RAINWATER IN THE SOIL

While the Howells were setting up their rooftop rainwater collection system they were simultaneously working in the landscape to harvest runoff water to serve their plants.

Dan and Karen were using all their roof runoff for their domestic water needs so they decided to set up a large food garden where the landscape naturally concentrated runoff water within a small arroyo (drainage) about 150 feet from the trailer and house. To stabilize the arroyo and ensure the gardens would be secure from erosive stormwater, Dan and Karen constructed a series of check dams and gabions in the drainage. Check dams are permeable barriers placed *within* a drainage perpendicular to the flow of water. Rock gabions are further stabilized with a wire fence wrapping.

The gabions and check dams were situated within the drainage so that once detritus and sediment backed up behind the small structure, they would create a series of level terraces stepping down from the top to the bottom of the arroyo. The flow of water was spread over a wide surface, slowing it down and allowing it to gently sink into the soil. Other earthworks slowed water flow further upslope within the local watershed draining to the arroyo. With erosion checked and water harvested in the soil, new vegetation started to appear in the once-deteriorating arroyo bed. As Dan points out, "If you have scouring in an arroyo and no vegetation, you know the situation is out of control. If vegetation is established on the bottom you know the situation is more stabilized."

Within these stabilized level terraces Dan and Karen planted 600 square feet of gardens meeting 15–25% of their food needs. They had the most success with asparagus, garlic, Egyptian walking onions, and Jerusalem artichokes. The gardens are watered primarily from rainwater stored in the soil, but for less drought-tolerant plants the Howells wanted a source of water for surface irrigation in dry times. To support this, they hand dug two 20,000-gallon dirt reservoirs high in their landscape. These reservoirs completely fill in one good rain, and provide all the irrigation water needed for an entire year! Once the reservoirs are full, water is pumped from the reservoirs to a 10,000-gallon fiberglass tank and a used 5,000-gallon steel tank also sitting high in the landscape. Water is then distributed from the tanks to the gardens using gravity flow.

To support water-needy vegetables, Dan and Karen arranged soaker hoses several inches below the land's surface and mulched the soil above to retain moisture. Ten pounds per square inch (psi) of water pressure is needed for the irrigation lines to function, and this is easily achieved by the tanks' placement above the elevation of the garden: *Each foot a water source (tank) is raised above its destination (garden), gravity provides 0.43 psi of pressure.* As Dan says, "No pumps, no utilities, and no pollution." (See the chapter on principles for cistern systems in volume 3 for a gravity-fed drip irrigation system that can work on less than 10 psi of pressure.)

After each rainy season surplus water in the reservoirs is siphoned out over the landscape where it's needed. Accumulated silts are removed from the bottom of the reservoir and wheel-barrowed downslope to make water-harvesting contour berms. Sometimes a

---

Box 3.2. The System Is Low-Maintenance, But Definitely Not Maintenance-Free.

Three years ago a broken pipe resulted in the irrigation system losing 7,000 gallons of harvested rainwater. Unlike a municipal system where there is a backup water supply, Dan and Karen's system is limited and leaks must be dealt with right away. They made it through the season, but they had to be extra conservative until the rains renewed their water storage. Frequent inspection of your system is a very good idea.

---

basin is dug along the upslope side of these berms, making a berm 'n basin. These structures are laid out in lines of equal elevation along the land, where they intercept and quickly infiltrate rainwater into the soil where it is available to vegetation.

Dan and Karen are always experimenting to try to increase the productivity of their site and the efficiency of their water harvesting. A rain-fed orchard within a net and pan system of berms had limited success. From this they discovered that the microwatersheds needed to be enlarged to harvest more water, and that climate extremes limited the diversity of productive crops they could grow. More successful was the creation of different styles of berm 'n basins that sped up their revegetation work (for more on the earthworks techniques mentioned in this and the above paragraph, see the chapter on berm 'n basins in volume 2).

The Howells' work isn't over yet. They have a life far from the 9-to-5 office day, but not a life of leisure. The system needs to be maintained, and they have plans to dig out more reservoirs and create more contour berms. They plan to use the additional harvested water to cultivate native and medicinal plants. When the storm clouds break over their land, berm 'n basins and tanks fill with water, gabions catch fertile silts, and vegetation bursts into new growth. More resources are now being generated on their site than drained away. It's been over twenty years since that fateful day in the California laundromat when Dan decided to run with a wild idea, who would've guessed it would have them both running with the rain.

Integrated Design

This chapter shows you how to maximize the potential of your site's water resources by integrating harvested water with sun and vegetation at your site to help passively cool buildings in summer, heat them in winter, and enhance plants and gardens. By doing so, you realize the seventh principle of water harvesting: Maximize beneficial relationships and efficiency by "stacking functions."

Use the strategy of "integrated design" to provide on-site needs (e.g., water, shelter, food, aesthetics) with on-site elements (e.g., stormwater runoff, greywater, cooling shade, warming sun, vegetation) by creating an efficient design that saves resources (e.g., energy, water, money) while improving the function and sustainability of the site.[1] Integrated design helps turn "problems" into solutions. For example, erosive floodwater runoff harvested into basins can provide water to grow shade trees, helping control flooding and erosion. The key is to *see, understand,* and *combine* on-site elements—such as stormwater runoff, vegetation, and solar exposure—to maximize their beneficial use.

To help you do just that, this chapter helps you develop an integrated design for your site. Once again we start with observation, this time focusing on the sun's path across your site, and mapping it along with the other observations you've plotted on the site plan you created in chapter 2. Why look at the sun in a water-harvesting book? That is so you can orient buildings, plantings, and more to maximize the degree

to which they can produce resources, rather than consume them—by passively and freely heating, cooling, powering, growing, and maintaining themselves in a way that will make all your water resources (and time and money) go further. In that vein, *seven basic patterns* of integrated design are presented in this chapter to help you create a conceptual layout of water-harvesting earthworks, tanks, gardens, trees, and buildings that work off your observations and build on your site's existing resources while helping mitigate its challenges. The more patterns you incorporate into your design, the more integrated it becomes. The chapter ends with tips on how to further refine your site's integrated conceptual design, and the story of how my brother and I created and implemented such a plan for our urban lot. (Appendix 5 provides worksheets prompting you for information to go along with the following seven principles. You may want to read this chapter straight through first, then read it with the worksheets, marking down appropriate information.)

THE PATH OF THE SUN

We live on a planet with a 23.5° tilt that travels completely around the sun each year. These characteristics result in a gradual shift throughout the year in the direction and time of day that the sun rises and sets, and the angle of the sun above the horizon. The

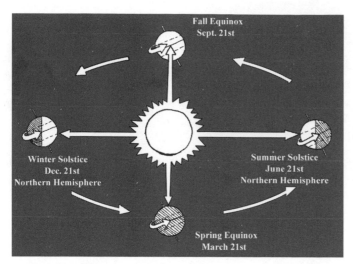

Fig. 4.1. Changing angles of the sun's rays cause the changing seasons. On the winter solstice in the northern hemisphere, the sun is low in the sky at noon, and it does not even appear at the north pole; but at the same time in the southern hemisphere, the sun is directly overhead at the Tropic of Capricorn. On the summer solstice in the northern hemisphere the sun is high in the sky at noon, the sun does not set at the north pole, and the noonday sun is directly overhead at the Tropic of Cancer. Adapted from *The Hand-Sculpted House—A Practical and Philosophical Guide to Building a Cob Cottage* by Ianto Evans, Michael G. Smith, and Linda Smiley, Chelsea Green Publishing, 2002

degree of shift depends upon your site's latitude on earth. (See figure 4.1.)

In the northern hemisphere, the sun rises *south* of due east and sets *south* of due west in the winter, while it rises and sets *north* of due east and west, respectively, in the summer.

In the southern hemisphere it is the opposite: The sun rises and sets north of due east and west in their

winter (which occurs while the northern hemisphere is having summer), and it rises and sets south of due east and west in their summer. On the spring and fall equinox (March 21 and September 21 in the northern hemisphere; September 21 and March 21 in the southern hemisphere) the sun rises and sets due east and west.

ORIENT YOURSELF TO THE SUN'S "FLOW" THROUGHOUT THE YEAR

For those living in the northern hemisphere, the south-facing side of buildings, walls, and trees is the "winter-sun side" and the north-facing side is the "winter-shade side." This is because the winter sun stays in the *southern* sky all day. Midday the sun's angle off the horizon remains low (the angle gets lower the further north you are in latitude). (See figure 4.2.)

It is the opposite in the southern hemisphere. The north-facing side of the buildings, walls and trees is the "winter-sun side" and the south-facing side is the "winter-shade side." The winter sun remains in the *northern* sky.

Identify the "winter-sun side," and the "winter-shade side" of your home now! The rest of this chapter continually refers to this orientation, so get ready and get oriented now!

In summer north of 23.5° N latitude, the sun rises and sets north of due east and west, but at midday the sun is in the southern sky (and higher off the horizon than in winter). So the "summer-shade side" of buildings, walls and trees is the south-facing side in the morning and late afternoon, but midday it is the north-facing side (with a much shorter shadow than is cast in winter). The converse is true in the southern

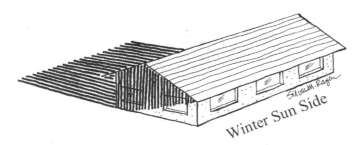

Fig. 4.2A. Winter sun exposure and shade cast at noon on the winter solstice at 32° latitude

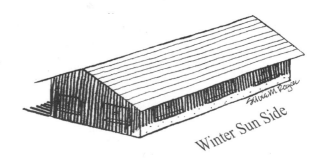

Fig. 4.2B. Summer sun exposure and shade cast at noon on the summer solstice at 32° latitude

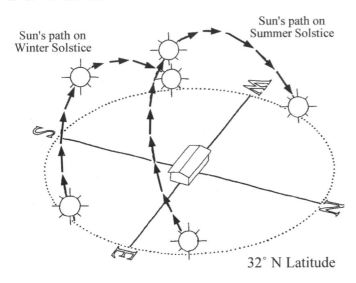

Sun's path on Winter Solstice

Sun's path on Summer Solstice

32° N Latitude

Fig. 4.3. The winter and summer solstice paths of the sun at 32° N latitude

hemisphere. (See figure 4.3; see also box 4.2 for sun angles by latitude and season, for the northern and southern hemispheres.)

Residents of the tropics (from 23.5° N and S latitude to the equator) have a simpler situation. The noonday summer solstice sun is more than 90° above the horizon (south horizon in the northern hemisphere; north horizon in the southern hemisphere), so the "summer-shade side" of various objects is their south-facing side (northern hemisphere) or their north-facing side (southern hemisphere) all day long.

Identify the "summer-shade side(s)" of your home now!

ADD THE LOCATION OR FLOW DIRECTION OF THE SUN AND OTHER OBSERVATIONS TO YOUR SITE PLAN

Map the location of the rising and setting sun on the summer and winter solstice; where you would like more shade or exposure to sun; the direction or location where prevailing winds, noise, or light come from; the foot traffic patterns of people, pets, or wildlife; and any other resources or challenges you may want to design for (fig. 4.4). By recording your site's existing resources and challenges you can improve the layout and design of water-harvesting earthworks, tanks, gardens, shade trees, paths, and

buildings so they harvest more resources and diffuse or divert the challenges.

THE SEVEN INTEGRATED DESIGN PATTERNS

Now use the following integrated design patterns to get ideas on how you can efficiently arrange elements of your design to the unique conditions and needs of your site. You will want to have your site map (and worksheets) handy on which to write any additional information. All "Action Steps" discoveries and calculations should be written on your worksheets for further reference. These patterns have a sequence, as you will see.

INTEGRATED DESIGN PATTERN ONE

Orienting Buildings and Landscapes to the Sun

Integrate the orientation of buildings, living spaces, and water-harvesting earthworks/planting areas

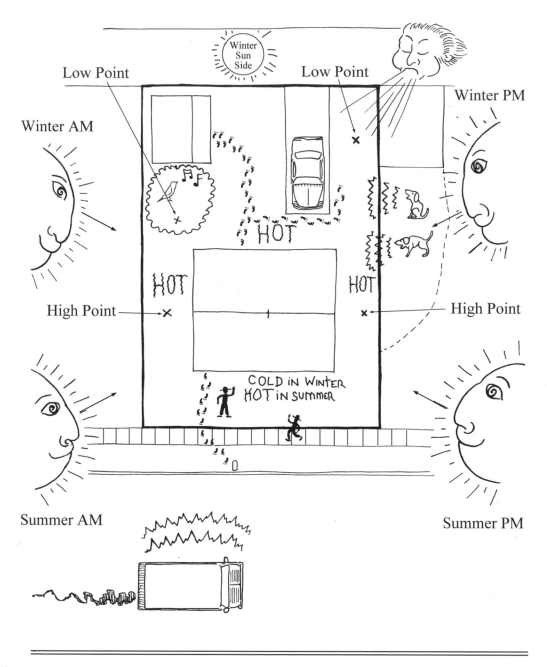

Low Point

Low Point

Winter AM

Winter PM

Winter Sun Side

HOT

HOT

HOT

High Point

High Point

COLD iN WINTER
HOT iN SUMMER

Summer AM

Summer PM

Fig. 4.4. Map the resources and challenges on your site, sample map. See appendix 5 for worksheets.

with the sun to maximize passive heating and cooling while reducing water and power needs. A year-long study in Davis, California, monitored temperatures in two identical apartment buildings with different orientations to the sun. No heating or cooling systems were operated during this year. The study found that apartment units in the building *with an east-west orientation* (long walls facing south and north) *and with* *small roof overhangs were 17°F warmer in winter and 24°F cooler in summer* than apartments in a similar building with a north-south orientation (long walls facing east and west).[2] (See figure 4.5.) That is a huge difference! Building or buying a home with correct solar orientation costs nothing extra, yet it can drastically reduce utility costs by maximizing winter sun warmth and minimizing summer heat.

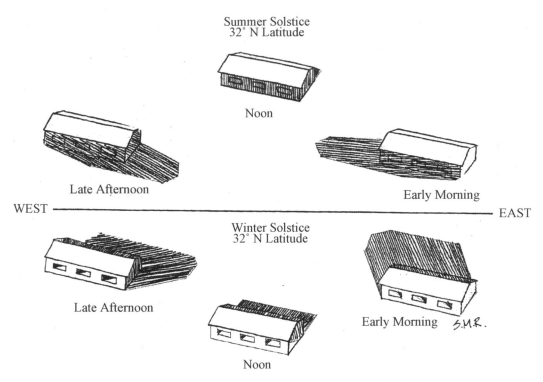

Fig. 4.5A. Sun exposure on, and shade cast by, an east-west oriented building at 32° N latitude. View of building's winter-sun side. This orientation is warmer in winter and cooler in summer. Note how the winter-sun side is shaded by the roof overhang at the summer solstice, but not the winter solstice.

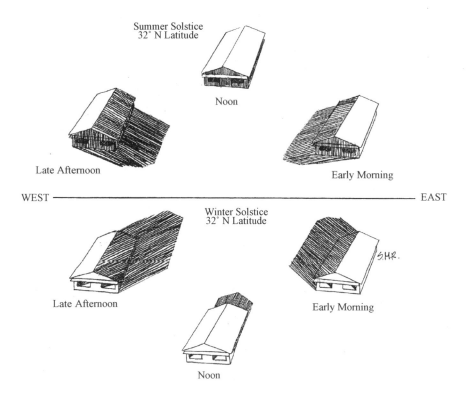

Fig. 4.5B. Sun exposure on, and shade cast by, a north-south oriented building at 32° N latitude. View of building's winter-sun side. This orientation is colder in winter and hotter in summer.

Box 4.2. Approximate Sun Angles by Latitude for Northern Hemisphere in Summer (June 21) and Winter (December 21), and for Southern Hemisphere in Winter (June 21) and Summer (December 21)

Find the latitude of your site by looking at a globe, atlas, or topographic map, or google "What is the latitude of (your town, state, and country)?" Note: The angles in this table have been rounded off for easier reading. See appendix 6, section L for more resources on sun angles and passive solar design.

Latitude, °N or S	Date	Location sun rises, N or S of due East	Location sun sets, N or S of due West	Northern hemisphere: Location of noonday sun, above southern horizon	Southern hemisphere: Location of noonday sun, above northern horizon
20°	June 21	22° N	22° N	93°	47°
	Dec. 21	22° S	22° S	47°	93°
24°	June 21	24° N	24° N	89°	43°
	Dec. 21	24° S	24° S	43°	89°
28°	June 21	26° N	26° N	85°	39°
	Dec. 21	26° S	26° S	39°	85°
32°	June 21	28° N	28° N	81°	35°
	Dec. 21	28° S	28° S	35°	81°
36°	June 21	30° N	30° N	77°	31°
	Dec. 21	30° S	30° S	31°	77°
40°	June 21	32° N	32° N	73°	27°
	Dec. 21	32° S	32° S	27°	73°
44°	June 21	34° N	34° N	69°	23°
	Dec. 21	34° S	34° S	23°	69°
48°	June 21	36° N	36° N	65°	19°
	Dec. 21	36° S	36° S	19°	65°

The buildings and trees in your landscape cast cooling shade during the day and the vegetation's canopy reduces radiant heat loss at night creating a diverse array of microclimates. When planting within water-harvesting earthworks identify these microclimates, and select and place vegetation appropriate to these microclimates. Cold-sensitive plants go on the warm winter-sun side of a tree or building. Hardy drought- and heat-tolerant plants go on the west side where afternoon sun is hottest and evapotranspiration is greatest. Cold-tolerant plants go on the cool winter-shade side. Vegetation needing more water goes on the east side where plants will get sun on cool mornings and shade on hot afternoons.

Action Steps

• How is your site and/or your home oriented? If you don't know, get a compass or ask the sun. The sun will orient you to the cardinal directions if you use the information in box 4.2 and you pay attention to where the sun is throughout the day. Put this information on your site map.

• When you know your building's orientation, and have a good grasp of where north, south, east, and west are in relation to your site, move on to the following patterns. They will help improve the performance of your home and landscape, even if their orientation is less than ideal.

INTEGRATED DESIGN PATTERN TWO

Overhangs on Buildings

Like a broad-brimmed sun hat, roof overhangs on a building can improve your comfort. Roof overhangs can improve a building's passive cooling and heating performance regardless of orientation and increase roof area resulting in more roof runoff. Properly sized overhangs on a building's "winter-sun" side let in low angle winter sun while blocking overhead summer sun. Overhangs on the east, west, and winter-shade side of buildings boost summer shading. Combine overhangs with correct solar orientation and you'll be even more comfortable and pay far less utilities. (See figure 4.6.)

Given your latitude, how far should window overhangs extend out from a building's *winter-sun side*? The following equation from *The Passive Solar Energy Book* by Edward Mazria[3] provides an answer. (Note that the equation only applies to winter-sun-side-facing windows.)

$$\text{window height} \div F = \text{Overhang Projection (OP)}$$

Window height is determined by measuring the vertical distance from the windowsill to the bottom height of the overhang's extension (see figure 4.6), while F is a factor selected from the table in box 4.3 according to your latitude and climate. (Note: If you will be installing gutters, be sure to include their width as part of the overhang's extension, since they will extend the overhang and the shadow it casts.)

To determine the ideal sizing of *windows* in different climates to enhance passive heating and cooling potential, see *Sun, Wind, and Light* by G. Z. Brown and Mark DeKay.

Overhangs for Gardens

Here in the low desert of southern Arizona, living overhangs can provide the winter sun and diffuse summer shade vegetable gardens crave. Place sunken garden beds on the south or southeast side of pruned-up native mesquite trees to allow winter sun to hit the garden directly. In summer overhanging branches will

shade out much of the overhead sun. I can produce salad greens, artichokes, herbs, snow peas, garlic and onions, potatoes, carrots, Jerusalem artichokes, and edible flowers in full winter sun with mild temperatures and low evaporation rates. In the extreme heat of summer when evaporation rates increase, the diffuse shade of the mesquite keeps chiles, tomatoes, basil, eggplant, squash, gourds, and summer greens from prematurely wilting from exposure.

Action Steps

- Shut off mechanical heating and cooling systems once in each season of the year to observe how direct solar exposure—or the lack of it—affects the comfort of your home.

- Use the overhang projection calculation to determine the appropriate overhang sizes for your area. Compare the existing overhangs to what the calculation recommends. Observe how the overhangs affect your comfort.

Box 4.3. Latitude and F factor

North or South Latitude	F factor
28°	5.6–11.1
32°	4.0–6.3
36°	3.0–4.5
40°	2.5–3.4
44°	2.0–2.7
48°	1.7–2.2
52°	1.5–1.8
56°	1.3–1.5

Using lower F factors provides more shade for more of the summer. Those living in a climate of hot summers and mild winters will likely want to use the lower number in the F factor range when calculating overhang length; those living in climates of mild summers and cold winters will likely want to use the higher number in the F factor range.

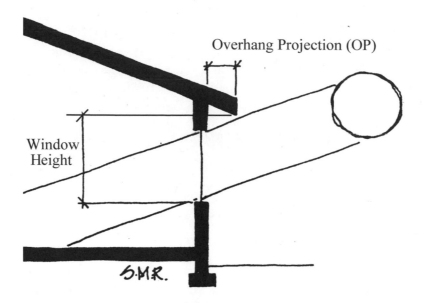

Overhang Projection (OP)

Window
Height

S.M.R.

Noon at Winter Solstice

Fig. 4.6A. A winter-sun-side roof overhang allowing winter sun exposure for a window at 32° latitude

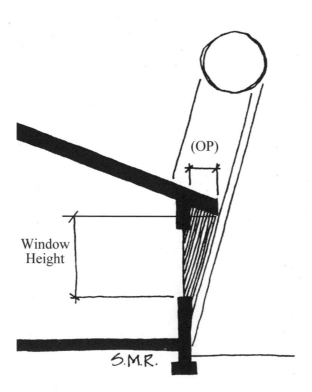

(OP)

Window
Height

S.M.R.

Noon at Summer Solstice

Fig. 4.6B. A winter-sun-side roof overhang providing summer shade for a window at 32° latitude

• When installing or extending an overhang, consider the following options:

- An overhang extending across the whole winter-sun-side of your building to benefit your wall as well as your windows.

- Awnings just for your windows (fig. 4.7).

- A very sparse trellis to support dense winter-deciduous vines which provide leafy shade in summer. (See figure 4.8.) In the winter, remove and mulch the uppermost sections of the leafless vines so your windows get direct winter sun (see "Integrated Design Pattern Five: Maintaining Winter Sun Exposure"). Use rooftop runoff to irrigate the vines. (Note: While water-harvesting earthworks should be placed a minimum 10 feet from the house, vegetation can be planted closer to a building. Just train the roots

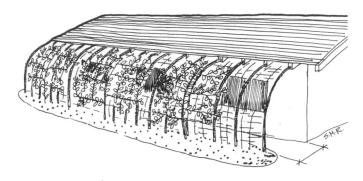

Fig. 4.8. Winter-sun-side trellis

Fig. 4.9. East-facing exterior blinds doubling as pollinator habitat. Beneficial carpenter bees live in the hesperaloe flower stalks of which the blinds are made. Note dappled shade from trees to east provides additional cooling!

Fig. 4.7. Window awning

of the plants to find the water harvested in the more distant earthworks. Do this by applying irrigation water on the side of the plant closest to the earthworks. Then every month or two move your irrigation emitter or hose a foot further from the plant and a foot closer to the earthworks, until you eventually end up watering the plant via its extended roots within the earthworks.)

- Exterior blinds or shutters on east- and west-facing windows to help block direct summer sun before it enters and heats the building (fig. 4.9).

- Covered porches and/or shade trees on the east, west, and summer-sun sides of your building as presented in the next pattern.

INTEGRATED DESIGN PATTERN THREE

The Solar Arc

A solar arc is created using a number of shading elements such as trees, cisterns, trellises, covered porches, and overhangs laid out in the shape of a big open-armed hug that welcomes the full potential of the winter sun (fig. 4.10). At the same time, it deflects much of the summer sun using the arc's "back, shoulders, arms, and hands." Water-harvesting earthworks are the foundation of solar arcs when vegetation is used as the sheltering element (figure 4.11 shows the growth of the solar arc from basin to young trees to mature trees). Situate the earthworks and trees close enough to buildings to use roof runoff as the primary source of irrigation water and household greywater as the secondary source. As the shade trees grow they beautify your yard, clean the air, and dramatically cool summer temperatures (see boxes 4.4 and 4.5).

Solar arcs shade buildings, gardens, and gathering spots in the yard from the summer's northeastern morning and southwestern afternoon sun. Put trees that drop their leaves in winter on the east and west arms of the arc to allow filtered sun to penetrate the arc early in spring and late into fall. Evergreen trees work well on the northern band of the arc to block summer sun and deflect cold northerly winter wind. The same principle works in the southern hemisphere, only the directions change.

Box 4.4. Growing Your Air Conditioner

A study conducted in Phoenix, Arizona found that water use in evaporative coolers averages 65 gallons (245 liters) per cooler per day, or about 13,400 gallons (50,900 liters) during the cooling season from March to October.[4] That same amount of water could fulfill all the water needs of four native mesquite trees with 20-foot (6-meter) heights and canopies.[5] If placed on the east, west, northeast, and northwest sides of a home, these shade trees could reduce summer temperatures around the building by as much as 20°F compared to the same building without shade.[6]

Box 4.5. "Cool and Clean" or "Cool and Polluted"

The generation of electricity used to mechanically air condition an average household causes about 3,500 pounds of carbon dioxide and 31 pounds of sulfur dioxide to be released from powerplant smoke stacks each year.[7] Cooling solar arcs of trees consume carbon dioxide and produce oxygen—up to 5 pounds of oxygen per day per tree.[8] According to the National Arbor Day Foundation, over a 50-year period, a well-placed shade tree can generate $31,250 worth of oxygen and provide $62,000 worth of air pollution control.[9]

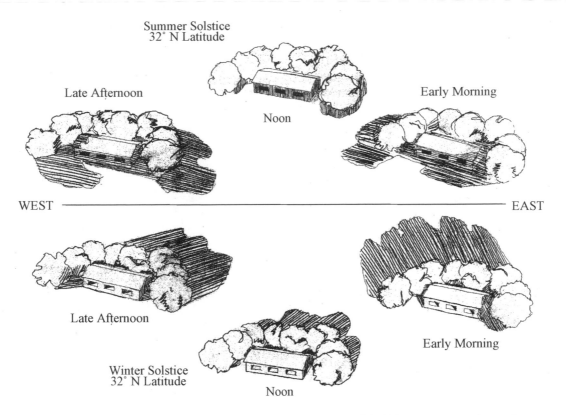

Fig. 4.10A. Solar arc of trees with an east-west oriented building at 32° N latitude. View of building's winter-sun side. Compare to figure 4.5A.

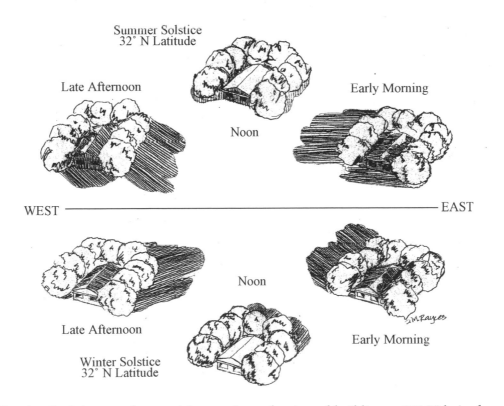

Fig. 4.10B. Solar arc of trees with a north-south oriented building at 32° N latitude. View of building's winter-sun side. Compare to figure 4.5B.

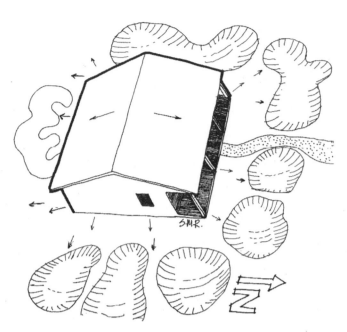

Fig. 4.11A. Water-harvesting basins placed to help grow a solar arc of trees, and one basin on the winter-sun side of the house to grow a sunken winter garden (32° latitude)

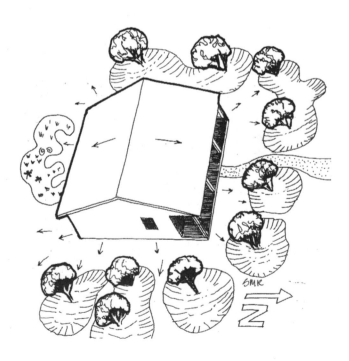

Fig. 4.11B. Young trees and the garden planted within the basins and irrigated with harvested roof runoff

Fig. 4.11C. Trees at full size forming a living solar arc, and a thriving winter garden

Action Steps

• See if you have any elements of a solar arc in place around your home or garden, such as an existing shade tree or building.

• Map where missing pieces of a solar arc should be located to complete it and benefit your home or garden. Create water-harvesting earthworks and/or install cisterns to sustain shade trees.

INTEGRATED DESIGN PATTERN FOUR

Sun Traps

A sun trap creates a nice place to plant a garden or take a nap. Elements making up a sun trap can include cisterns, tall low-water-use vegetation, a house, trellis, shed, or other shading elements. A sun trap is more open than a solar arc. In the northern hemisphere the L-shaped sun trap is open to the east and south, and closed to the north and west. In the southern hemisphere it is open to the east and north, while closed to the south and west. (See figure 4.12.) The sun trap creates microclimates ideal for gardens, sensitive plants, cozy outdoor gathering areas, and hammock roosts.

Fig. 4.12A. House, cistern, and tree in a water-harvesting basin forming a sun trap harvesting the afternoon sun in *winter*. You are looking at the winter-sun (south) side of the house in the afternoon—the sun setting south of due west.

Fig. 4.12B. The same house, cistern, and tree in a water-harvesting basin forming a sun trap deflecting the afternoon sun in *summer*. You are looking at the south-facing side of the house—the sun setting north of due west.

Fig. 4.13. The prevailing wind's view of Allegra Ahlquist's cistern and wall, creating a protective sun trap and windbreak for her courtyard. Solar panels on roof faces south.

Box 4.6. A Sun Trap Acting as a Windbreak, Fire Break, Privacy Screen, and Garden Courtyard

Dynamo grandmother Allegra Ahlquist lives in a cement-plastered sandbag house in the hot, windswept, fire-prone grasslands of southeastern Arizona. The environment seemed too harsh for gardening, much less sitting outdoors, so she built a cistern and garden wall southwest of her home's south-facing wall to create a sun trap. Now this sheltered nook creates a pleasant environment for year-round outdoor sitting and for a sunken-bed garden. The cistern harvests pure rainwater close to where it's needed in the garden, and shelters the vegetables from harsh westerly winds and hot afternoon sun. An outdoor shower, served with biocompatible soaps, drains greywater to perennial garden plants. (See figure 4.13.)

My sun trap captures winter sun in the mornings to burn off frosts and warm things up, but shades out sun on hot afternoons. This extends the cool growing season of my low desert garden by two months. I can plant a month earlier and get an extra month of growth by keeping my vegetables from bolting (going to seed). The afternoon shade reduces plant evapotranspiration, heat stress and drought stress, and reduces pest problems. I enjoy hanging out in my sun trap garden, because, like the plants, I get warmth in the morning when I want it, and shade in the afternoon when I need a break from the sun.

Action Steps

• Before you plant, identify and map the areas of your site where a sun trap might make sense, and map any existing elements already in place.

• Harvest rainwater to support vegetative elements in the sun trap. As Mr. Phiri says, "Plant the water before you plant the trees."

• Determine if a new cistern could be placed to help create a sun trap for a garden or patio.

INTEGRATED DESIGN PATTERN FIVE

Maintaining Winter Sun Exposure

As you harness the full potential of rain, do the same for the winter sun. Keep the exposure open for winter sun-facing windows, gardens, solar water heaters, solar panels, and solar ovens. The Village Homes development in Davis, California, found that with winter solar access retained, simple solar homes can achieve 40 to 50% of their winter heating needs from the sun, and more sophisticated designs can meet 85% of their heating needs.[10] I rely on south-facing windows for much of my home's heating needs; a solar water heater provides all hot water; and eight solar panels provide all of the power needs. But misplaced trees could seriously cripple this performance; even the shadow cast by a leafless tree could block out over 50% of the potential heat and light.[11]

When planning where to place cisterns or trees, and the water-harvesting earthworks that will sustain them, be conscious of the shadows they will cast. Put them where they won't block the winter sun-side of your home, solar water heaters, or winter garden. (See figure 4.14 for an example.) You can determine the longest shadow an object will cast on the winter solstice (December 21 in the northern hemisphere, June 21 in the southern) by looking up your latitude in box 4.7 and multiplying the object's height by the associated factor of the shadow ratio.

For example, here in Tucson at 32° N latitude, the ratio is 1:1.49, so for every foot of an object's height, the shadow cast at noon on December 21 will be 1.49 feet long. Multiply the height of a mature 25-foot tall tree by 1.49 to get 37.25—the length of the shadow (in feet) cast to the north at my latitude. Using this calculation I can determine how far south of my home or winter garden I need to place that tree so it won't block winter solar gain once it grows to mature size.

Action Steps

• Use the shadow ratio in box 4.7 to predict how existing vegetation and structures might affect your winter solar gain. Then observe what happens.

Box 4.7. Noontime Winter Solstice Shadow Ratios

Adapted from: *Effective Shading with Landscape Trees*, by William B. Miller and Charles M. Sacamano, University of Arizona College of Agriculture, Cooperative Extension bulletin 188035/8835, March 1990

North or South Latitude	Object Height: Length of Shadow Cast at Noon
28°	1:1.28
32°	1:1.49
34°	1:1.55
36°	1:1.75
40°	1:2.04
44°	1:2.50
48°	1:3.13
52°	1:3.70
56°	1:5.26

Box 4.8. Solar Rights

In 1977, the State of New Mexico enacted the Solar Rights Act, the first law in the country recognizing the natural resource of solar energy as a property right. The property owner who first claims solar rights can prevent neighboring property owners from encroaching upon the right with new buildings or trees. See appendix 6, section M for more on New Mexico's solar rights.

In Arizona, Solar Rights Law ARS 33-439 has banned any covenant, restriction, or condition from prohibiting the installation or use of a solar energy device.

The progressive communities of Village Homes in Davis, California and Milagro Cohousing in Tucson, Arizona state in their Declaration of Covenants, Conditions, and Restrictions that residents are prohibited from interfering with a neighbor's solar rights. A homeowner is, however, allowed to infringe on his or her own solar rights.[12,13] And rather than banning the use of clothes lines as some CCRs do, they encourage their use.

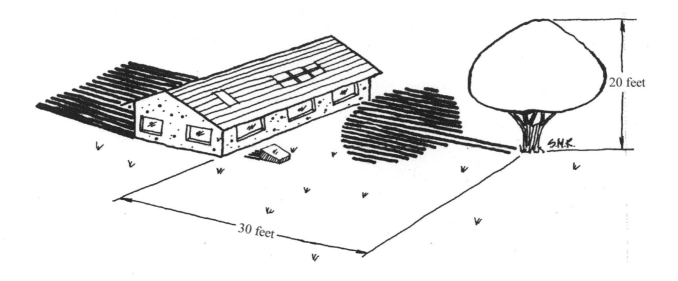

Fig. 4.14A. At 32° latitude a tree selected and placed, so at mature size, winter sun exposure (during the winter solstice at noon) will be maintained for a home's winter-sun facing windows, solar water heater, solar panels, and solar oven

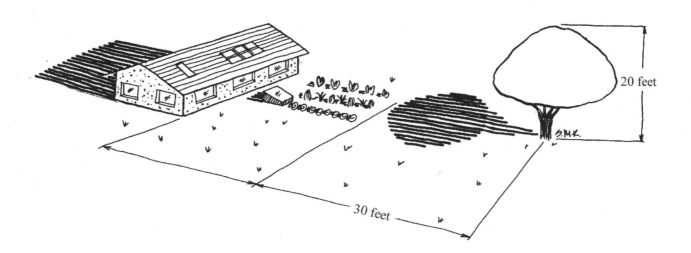

Fig. 4.14B. At 32° latitude a tree selected and placed, so at mature size, winter sun exposure (during the winter solstice at noon) will be maintained for a winter garden and the home, as well as the windows, solar water heater, solar panels, and solar oven

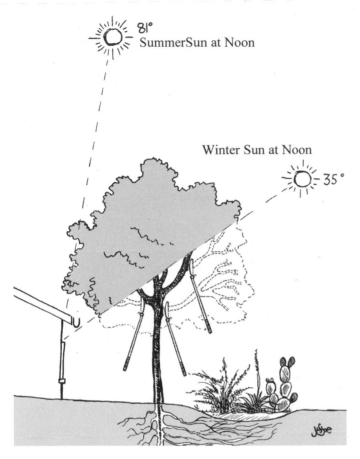

81°
SummerSun at Noon

Winter Sun at Noon

35°

Fig. 4.15. A tree pruned to allow direct winter sun access through winter-sun side windows, while shading out the summer sun. Adapted from *Designing and Maintaining Your Edible Landscape—Naturally* by Robert Kourik, Metamorphic Press, 1986

Box 4.9. Pruning for Winter Sun

If you currently have mature trees blocking your winter heat and light you can prune them to regain winter solar access. Use box 4.2 to determine what part of the sun's path you could reopen with pruning, then prune if appropriate; see also figure 4.15.

• Use the shadow ratio calculation to correctly place new structures or vegetation on the winter-sun side of your buildings or winter garden. Plan for plants' sizes at maturity, not the size at planting time. Consult nursery staff and plant books for size information.

• Plan water-harvesting earthworks and cistern systems to support correctly placed vegetation.

INTEGRATED DESIGN PATTERN SIX

Raised Paths, Sunken Basins

Keep access ways "high and dry" and planted areas "sunken and moist." Always pair a raised path with a sunken basin to capture runoff and grow shelter and beauty for the path (fig. 4.16). Trees planted in these water-harvesting basins shade and beautify the adjacent roadways, paths, and patios. This reduces excessive solar exposure and in turn the risk of skin cancer, the fastest growing form of cancer in the United States,[14,15] while creating a comfortable place to drive, walk, ride a bike, or converse.

Fig. 4.16. Raise pathways, and sink mulched and vegetated basins.

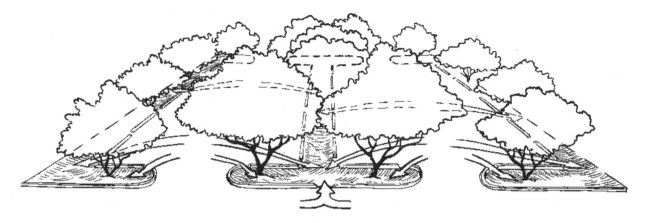

Fig. 4.17. A parking orchard of low-water-use, food-producing shade trees passively irrigated by runoff from a raised parking area harvested within sunken, mulched basins

Shade trees absorb much of the rain falling within the diameter of the tree canopy and runoff flowing around their bases. This creates a living flood control system and filters runoff-borne contaminants like nitrates, phosphorus, and potassium, which trees consider food.[16]

A parking lot can be planted with a "living carport" of flood-controlling, pollutant-filtering native shade trees irrigated solely by the parking lot's runoff. If low-water-use food-producing trees such as the velvet mesquite are planted, it becomes a "parking orchard." (See figure 4.17.)

Action Steps

• Observe the relative height of paths, sidewalks, driveways and streets compared to adjacent planting areas in your home and community. Do you see the "raised path, sunken basin" pattern or a sunken path, raised planting area pattern? Is stormwater being directed to vegetation, asphalt, or storm drains?

• Observe undisturbed natural areas. You will most likely find the largest, densest vegetation in depressed areas and along drainages where water concentrates.

• Identify and map areas where you can develop the raised path, sunken basin pattern at home. Create

water-harvesting earthworks by digging sunken basins, then use the newly available dirt to create the raised paths.

INTEGRATED DESIGN PATTERN SEVEN

Reduce Paving and Make It Permeable

Rainwater is like a naked person—it won't stick around if you put it on hot pavement in summer. So, we must reduce the amount of impervious paving on our sites and neighborhoods, while increasing shady vegetation (compare figures 4.18A and 4.18B).

Much of the heat stored in urban concrete and asphalt during the day is released in the late afternoon and evening, keeping temperatures high. Excessively wide, unshaded streets and dense unshaded development directly contribute to the heat island effect, and were found to raise maximum daytime temperatures by 10°F in Davis, California.[17] (See figure 4.19A for an example of a wide, tract-home street.)

Typical residential streets in the western U.S. are up to 40 feet wide, but there are alternatives. At the innovative Village Homes housing development in Davis, California, streets are 20 feet wide.[18] By narrowing streets, using cul-de-sacs, and limiting driveways to the length of a vehicle, Village Homes made 15% more land available for community gardens, orchards, tree-lined walkways, and bike routes.[19] Adjoining trees

Fig. 4.18A. A dehydrated and exposed residential lot dominated by impervious pavement, sparse vegetation, and bare compacted earth graded to drain all runoff to the street.

Fig. 4.18B. A hydrated and sheltered residential lot dominated by water-harvesting earthworks, native low-water-use vegetation, and permeable paving graded to minimize runoff and utilize it on-site.

Fig. 4.19A. A wide, exposed, solar-oven-like street in Tucson, Arizona

Fig. 4.19B. A narrow, mature tree-lined, and shaded street in Village Homes, Davis, California.
See Fig. 3.17 for a narrow, tree-lined street in Tucson, Arizona.

Fig. 4.20. Admiring the birds in the canopy of oak trees forming a living outdoor classroom at the Wildflower Center in Austin, Texas

create a canopy over the narrow streets, shading 80% of the streets' area, reducing summertime temperatures, reducing the need for buzzing air conditioners, and creating a much quieter and more pleasant neighborhood (fig. 4.19B).

Direct street runoff to adjoining water-harvesting earthworks, try to reduce the amount of paving, and see if remaining pavement can be made permeable. In volume 2, the chapter "Reducing Hardscape and Creating Permeable Paving" features a number of permeable pavements that can reduce runoff by up to 90% and yield significantly cleaner stormwater compared to impervious pavements.[20]

Action Steps

Look for examples of pervious and impervious paving around your home and community.

- Determine how you can reduce the paving on your site, and make remaining pavement more permeable. Implement these changes.

- Consider turning your driveway into a park-way by limiting its size to the length and width of your vehicle.

- Consider using porous brick, cobbles, or stabilized gravel in place of impervious paving materials.

- Plan water-harvesting earthworks and appropriate plantings in areas where paving is removed, and adjacent to areas where paving remains.

- Consider growing outdoor "rooms" of shade trees to replace or reduce the need for more buildings (fig. 4.20).

- In volume 2, read the chapter "Reducing Hardscape and Creating Permeable Paving."

INTEGRATING THE DESIGN ELEMENTS OF YOUR SITE TO CREATE A REGENERATIVE LANDSCAPE

The preceding integrated design elements should have you thinking about, mapping, and planning *where* you can lay out water-harvesting strategies to create a more integrated, efficient, and productive design. These elements are meant to give you a conceptual framework for creating "regenerative systems." Regenerative systems are created through thoughtful integration of design elements so they can enhance and recreate themselves without dependence on imported inputs such as piped-in fuel or pumped-in water. Once established, these regenerative systems do the bulk of the work, so you don't have to. They maximize the return on your water-harvesting investment. Anytime we invest effort, time, money, material, or labor we basically do so in one of three ways: **degeneratively**, **generatively**, or **regeneratively**. Compare their characteristics:

A **degenerative** investment:
- Starts to degrade or break down as soon as it is made;
- Requires ongoing investments of energy and outside inputs to keep it functional;
- Consumes more resources than it produces;
- Typically serves only one function.
- *Examples include*: ornamental lawns and landscapes dependent on chemical pesticides, fertilizers, and irrigation water imported from deep wells or municipal utilities; mechanically heated and cooled buildings powered by imported energy; and conventional single-use parking lots.

A **generative** investment:
- Starts to degrade as soon as it is made;
- Requires ongoing investments of energy and outside inputs to keep it functional;
- Produces more resources than it consumes;
- Typically serves multiple functions.
- *Examples include*: multi-use landscapes (producing multiple resources such as food, beauty, and wildlife habitat); passively heated, cooled, and lit buildings; durable alternative energy products

such as solar, micro-hydro, and wind power systems (turning buildings into "clean" energy producers); parking lots that grow a carport orchard of food-producing shade trees using harvested stormwater; and constructed rainwater-harvesting structures that increase the use and accessibility of on-site water resources.

A **regenerative** investment:
- Can repair and recreate or regenerate itself;
- Starts to grow or improve once it is made;
- Does not require ongoing investments of energy and outside inputs to keep it functional;
- Produces more resources than it consumes;
- Typically serves multiple functions;
- Can reproduce itself.
- *Examples include*: multi-use landscapes living solely off natural rainfall and requiring no additional outside resources after establishment; self-regenerating natural forests and ecosystems; revolving community loan funds; and vegetative rainwater-harvesting structures that build and repair themselves after establishment.

Strive to make all your water-harvesting endeavors regenerative, and as the water farmer Mr. Phiri would say, "You'll be rhyming with nature." You may not get there right away, but just by passively harvesting water the way this book suggests, you'll rise from the degenerative to the generative level.

TYING IT ALL TOGETHER: CREATING AN INTEGRATED DESIGN

Use your site plan, with its mapped resources and challenges, as the foundation for selecting and placing water-harvesting structures to create an integrated design that increases site efficiency and maximizes site potential. To do so:

1. Make multiple copies of your site plan (drawn to scale) to use as base maps for draft observations and ideas.

2. Play with different conceptual water-harvesting plan layouts. I recommend two options:

continued on p. 102

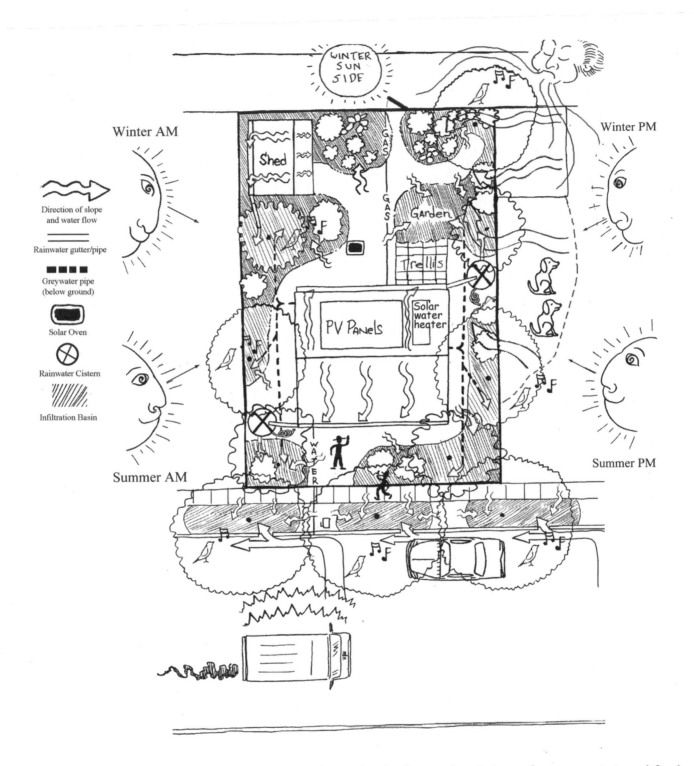

Fig. 4.21. An integrated rainwater- and greywater-harvesting landscape plan. Rain- and greywater-irrigated food-producing shade trees placed in a cooling solar arc around the home. Winter solar heating, power, and light retained. Impermeable hardscape reduced by removing driveway, planting a garden, and parking car on street. Garden placed in a sheltered sun trap and irrigated by a cistern forming part of the trap. A wind- and noise-break of hardy, low-water-use, song-bird-attracting vegetation placed within basins on the perimeter to diffuse and cool the wind. Pathways are raised to direct runoff to adjoining vegetation in basins. Roofs are guttered and sections of the land are regraded or bermed to utilize runoff flowing from off-site, and to retain all on-site rainwater on-site. Street and side-walk-runoff irrigated shade trees buffering home from street, and inviting friends to walk over and visit.

- Make cut-outs (in the same scale as your site plan) of the trees, cisterns, patios, gardens, and other elements you want to add to your site. Move these cut-outs around your site plan imagining how they will interact with the flows of your on-site resources (rainwater, greywater, sun, wind, etc.).
- Put tracing paper over your site plan and sketch where you could place various elements (trees, cisterns, patios, gardens, and other elements) you want to introduce to your site, and then see how they interact with on-site resource flows.

As you keep playing with various arrangements, ask yourself, "Where do I need water, where do I have it, how much do I have, and how/where can I best utilize it?" Remember, your goal is to increase site efficiency and maximize site potential. Refer to earlier sections of this chapter for conceptual ideas. See appendix 6, section N for information on additional design patterns.

3. *Refine your design by planning the water-harvesting details.* After figuring out where you want to harvest water, it's time to figure out how to do so. Re-read chapter 3 to determine what specific water-harvesting strategies are most appropriate for your needs, whether to harvest water in soil, tanks, or both.

Then go on to volume 2, which first provides an overview of water-harvesting earthwork strategies and their appropriate use, with the remaining chapters describing the specific water-harvesting techniques in detail, along with the use of vegetation, and greywater use in irrigating vegetation.

Then go on to volume 3, which describes cisterns. And now, walk your land again, imagining how various strategies could work within the unique context of your site. Play more with ideas and layouts on paper—it's much easier to make changes with a pencil and eraser than with a shovel (see figure 4.21 for an idealized site plan). When you feel you've got your plan set, scratch out, stake, or spray paint locations of paths, trees, water-harvesting strategies, and other elements in the dirt at your site. Walk around your site, feeling what it's like to inhabit this system. Make any needed changes and if all feels good—go for it!

REAL LIFE EXAMPLE

AN INTEGRATED URBAN RAINWATER-HARVESTING RETROFIT, TUCSON, AZ

Right after Rodd and I bought our east-west-oriented, fixer-upper of a house, the summer rains poured from the sky. We saw where the roof leaked, where runoff pooled against the house, and how the bulk of the rain ran off our site into the street. We mapped these observations, and others (noise, headlights, and pollution coming from street; where we wanted privacy; where we needed shade; and where we needed to enhance solar exposure to south-facing windows, etc.) on a plan we made of the property. More observations were added as we spent more time on the site, and we continue to do so today. Many hours were also spent imagining how we could improve the site with earthworks, cisterns, plantings, solar technologies, passive heating and cooling strategies, and more. We brainstormed how such improvements could be integrated with existing on-site needs and resources for maximum beneficial effect. (See figure 4.22, "before" and "after.")

Expanding on this practice of *long and thoughtful observation* we calculated the rainwater resources we could harvest within our site's watershed. In an average year of 12 inches (304 mm) of rain, about 6,000 gallons (22,800 liters) runs off our 990 square foot (91 square meter) roof, while an additional 38,000 gallons (143,600 liters) falls on our 132- × 46-foot yard. Just outside our fence 29,000 gallons (109,600 liters) of rain falls on the 20-foot (6-meter) wide public right-of-way on the south and east sides of our corner lot, while another 30,000 gallons (113,400 liters) per year can be harvested off the adjoining residential streets. In addition, a 270-square-foot section of our neighbor's roof drains 1,600 gallons of runoff into our yard. This totaled about 104,600 gallons (397,000 liters) of rain!

We *started harvesting roof runoff at the top of our watershed*—the roof. We removed a leaky asphalt roof and hauled the toxic pile of old asphalt to the dump—we didn't know then that asphalt could be recycled. We expanded the roof area by calculating and installing an extension of our roof overhang just

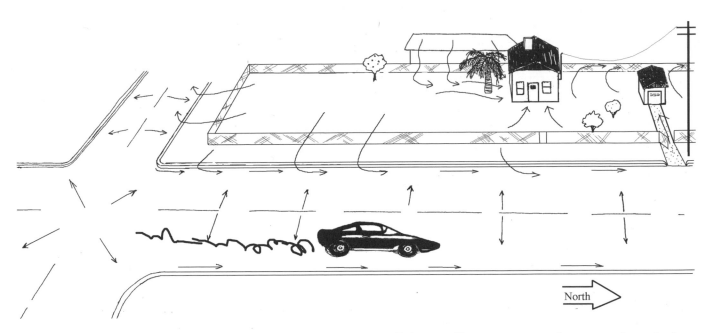

Fig. 4.22A. Our site at time of purchase in 1994. Most runoff drains off site, up against house, or through the garage. All greywater goes down the sewer. Palm tree blocks winter solar access.

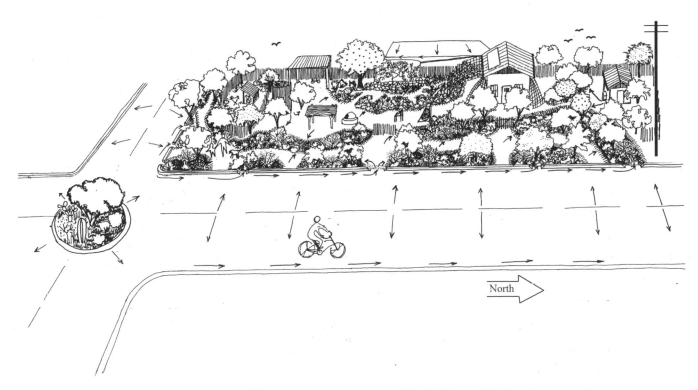

Fig. 4.22B. Our site in 2005. No runoff leaves site. On-site runoff is infiltrated before it gets to house or garage, and we have positive drainage away from buildings. Street runoff is directed to basins and trees along the curb. All greywater is directed to and recycled within the landscape. With palm tree removed, winter solar access is regained. Solar panels have been installed on roof, and a solar oven and a solar hot water heater have been installed on the ground south of the south-facing trellis. (The solar hot water heater is on the ground because our old roof was not strong enough to hold the heater's weight.)

long enough for more summer shade, while still short enough to allow the free heat and light of the winter sun to enter our south-facing windows. However, we forgot to include the width of the gutter (we'd later install) when determining the ideal length of roof overhang so now (with the gutter installed) a bit more of the winter sun is shaded than would be ideal. Then we installed 26-gauge galvanized steel metal roofing that should last for the rest of our lives. Metal was selected for its durability, ease of installation, strength, and, above all else, its non-toxic nature, allowing for more uses of its higher quality runoff water.

We began harvesting the majority of the rainfall as close as possible to where it falls—in the landscape with *small and simple* earthworks. Basins were dug and well-mulched to harvest and infiltrate rainfall and runoff throughout our watershed, *starting at the high points of the yard and working down to the low points.* Overflow water was directed from the upper basins to the lower basins, which spread and infiltrated the water still more. The dug-out soil formed a network of raised paths and raised gathering areas. We focused the earthworks and the water where we wanted to stack functions with *multi-use* vegetation, forming a solar arc of cooling shade trees on the east, north, and west sides of our home; along the property line to create a living fence of native plants, beauty, and wildlife habitat; and around an existing citrus tree that would form part of a sun trap shading a future organic garden from the afternoon sun. The 200-square-foot (18-m²) garden of sunken, mulched basins was placed just south of the home since the low-growing veggies would not block winter sun from the south-facing windows. But first we had to remove a palm tree that was blocking the winter sun. With the palm removed we had ideal winter solar exposure for winter gardening (the most productive and least water-consumptive season in Tucson), passive solar heating of our home and water, and the potential to put solar panels on our roof to produce our own electricity. Between the garden and home we built a sparse trellis on which we grow winter deciduous vines for summer shade. In winter we cut back the vines to let the sun shine in. More summer shade is created with a ramada and trees south of the home and garden, yet all are far enough to the south to maintain the winter's beneficial solar gain where we want it. The

sun is now our home's main source of heat, powers our solar oven, helps grow an abundance of winter greens, and with the installation of solar panels and a solar hot water heater, is our sole source of electricity and heat for water. Our home's east-west orientation, extended overhang, solar arc of shading vegetation, and other passive strategies such as nighttime ventilation are our main source of cooling. Electric bills no longer exist, and our gas and water bills are for little more than the service charge, since consumption is negligible to non-existent. (See figure 4.23.)

Along with sun and water, we also harvest the power of gravity. We guttered the section of the neighbor's roof draining onto our property to redirect the runoff to the high point of our yard where the citrus tree is located. Water used to drain away from the tree, now it passively drains to it. Gutters and the pitch of our roof direct just under half of the roof's runoff to earthworks and fruit trees north of the house, and the rest is directed to an above-ground cistern installed west of the garden along our property boundary on top of a 2-foot (60-cm) high earthen platform. The cistern, coupled with the citrus tree, provides multiple functions by enhancing the beneficial microclimate of a sun trap for the garden, acting as part of the property fence, and provided a privacy screen from a peering neighbor. By elevating the cistern we can use gravity to move water from the roof's gutter to the tank, and from the tank to the garden. Gravity pressure is low, so having the garden right next to the cistern keeps our hose length to just 25 feet (7.6 m), reducing pressure-reducing friction inside the hose and making cistern water convenient to use (fig. 4.24).

The cistern has a 1,200-gallon (4,560 liter) capacity. We selected this size after calculating the average annual roof runoff, assessing our water needs, and determining the resources we wanted to commit to the system (discussed further in *Rainwater Harvesting for Drylands*, volume 3). We knew that we did not have enough runoff to meet both domestic water needs and garden and landscape irrigation needs, so we implemented conservation strategies such as installing a composting toilet, installing a greywater system recycling all the water going down our drains within the landscape (see the greywater chapter in *Rainwater Harvesting for Drylands*, volume 2, and also

Fig. 4.23. The winter-sun side of our home in late winter. The solar panels (on roof), solar hot water heater (on ground to right of house), homemade solar oven (in front of water heater), solar food dryer (wood frame and screen set on top of the rebar trellis), and winter garden are all placed within the same solar envelope of open access to the winter sun for the south-facing windows. We then strive to shade the rest of the site with trees where the winter solar exposure does not need to be maintained. Note how the gutter extends the overhang and casts more winter shadow than we intended.

figures 4.25 through 4.28), and replacing some water-hungry exotic plants with drought-tolerant natives. This helped a lot, but we still did not have enough roof runoff to meet all our water needs. So we committed to using the roof water as part of an experimental start-up cistern system just for irrigation and supplemental outdoor water use. Wanting to keep the tank affordable and under a 1,500-gallon (5,700-liter) capacity, we decided to store the volume of water from a large 3-inch (75-mm) rainfall event that would drain off the 650-square-foot (58-square-meter) section of roof sloping toward the garden. This meant we needed a 1,200-gallon (4,560-liter) capacity tank.

Knowing how large a tank we needed, the next question was: What kind of tank did we want to use? Locally available pre-manufactured tanks included metal, plastic, fiberglass, or precast septic tanks. We also had the option of making our own ferrocement tank, or culvert cistern. (For more on tank options see *Rainwater Harvesting for Drylands*, volume 3).

We opted for a precast septic tank for the following reasons:

1. In terms of locally available, ready-made tanks, it was our cheapest option for the volume we needed.

continued on page 108

Fig. 4.24. Late winter garden grown entirely from rainfall and roof runoff harvested in 1,200-gallon tank in 2005

Fig. 4.25. Three-way diverter valve under sink gives us the convenient option of sending greywater to the landscape or the sewer depending on what goes down the drain.

Fig. 4.26. Greywater drains (marked with destination: fig, white sapote, orange, and peach) beside washing machine. Drain hose from washer is placed in a different pipe with every load of laundry. An option for households occasionally using non-biocompatible detergents is to include an additional marked drainpipe going to the sewer.

Fig. 4.27. Each greywater drain outlet is placed in a mulched and vegetated infiltration basin that also harvests rainwater and runoff. The drain pipe empties into an upside down bucket with holes in the sides, creating a chamber of air into which roots will not grow, so the pipe will not clog. An example bucket chamber sits beside the basin so you have a better idea of what the bucket within the basin looks like. Ordinarily the flat rock beside the drain outlet is placed over the outlet to keep the greywater below the surface of the rock and mulch. The rock lining the basin stabilizes the basin's edge.

Fig. 4.28B. Siphon hand pump (from auto parts store) pulls tub greywater through tube. Once the water is flowing the pump is pulled off, the tube is placed beside the fig tree, and the siphon effect continues to draw the water. Buy the tube after the pump so you can size the tube diameter to fit the pump.

Fig. 4.28A. Vinyl tube suction-cupped to bottom of tub and run through window frame accesses greywater for landscape use and bypasses having to deal with otherwise inaccessible tub drain plumbing. Note that this type of system may not seem convenient, although it may be a useful option for otherwise inaccessible greywater.

Fig. 4.29. Food grown and processed on-site that was irrigated with rainwater

4. Concrete is relatively fireproof material so the tank could act as a firebreak in the event the neighbor to the west sparked a fire.

5. There was a frost-sensitive citrus tree just south of where we wanted to place the tank, so the mass of the tank could help moderate temperature fluctuations and create a slightly warmer microclimate for the tree on winter nights.

6. We wanted to play with the idea of using a septic tank as a cistern to get to know its faults and favors. We figured it could be a viable ready-made system we might end up installing for clients.

The septic tank was custom made for use as a cistern, and further reinforced for above-ground installation (*see Volume 3 for details*). The cost back then was $600, which included delivery and placement.

All has worked great, with 95% of our garden's irrigation water now provided from harvested rainfall (fig. 4.29).

I no longer feel that Rodd and I are living entirely out of balance with the water resources of our dryland environment. We no longer get all our water from overdrafted groundwater and water imported from distant watersheds. Instead we are shifting more and more to living within our rainwater budget and the natural limits of our local environment. We are not living entirely off rainfall, but we are reducing our degenerative dependence on our community's diminishing groundwater and imported Colorado River water.

Within our generative landscape, rainwater has become our primary water source, greywater has become our secondary water source, and groundwater is strictly an infrequently used supplemental source. Most of our established landscape has even become regenerative by thriving on rainwater alone. The further we go the easier and more fun it gets. It has become a game where we use our creativity to get more and more out of what we find within the natural limits of our site's watershed, while giving back more than we take. In that spirit, we set up the outdoor shower so the bather could either use pressurized municipal water at the shower head or cistern water distributed from a

2. The 5-foot-tall, 10-foot-long, 4-foot-wide septic tank was right for our space. We created an elevated pad to improve gravity-fed distribution of the stored water, but because the base of the house was considerably lower than the planned pad we could not have a tank taller than 6.5 feet or the top of the tank would be higher than the drip edge of our roof, eliminating the possibility of gravity-fed inflow into the tank.

3. We wanted to use the tank as part of our western property line fence, privacy screen, and western sunscreen. The 10-foot length of the tank worked well for these uses.

Fig. 4.30. Three drains at the bottom of the outdoor shower, each diverting greywater to different plantings in the landscape

Fig. 4.31A. Public right-of-way adjoining property with asphalt driveway freshly removed, 1994

Fig. 4.31B. Tree-lined footpath reviving the once sterile right-of-way, 2004.

shower bucket hung from a hook. (See figure 4.30 of outdoor shower drain.) All shower and household greywater goes directly to the landscape via subsurface bucket drains (see the greywater chapter in *Rainwater Harvesting for Drylands*, volume 2). Other strategies have included creating a greywater laundromat in our backyard, reducing impermeable hardscape by replacing our asphalt driveway with lush plantings within infiltration basins, working with neighbors and the City to replace 26% of the pavement from the corner intersection with a traffic circle planted with native vegetation, and harvesting street runoff within curbside mulched basins to grow a greenbelt of street trees along the public right-of-ways (see the chapter "Reducing Hardscape and Creating Permeable Paving"

in volume 2). Figure 4.31 shows "before" and "after" photos. As a result the landscape is growing and producing in abundance, and our utilities and cost of living are steadily dropping—not the water table. This was recently recognized when we won "First Place Homeowner Landscape under $10,000," "Best Water Harvesting," and the "J.D. DiMeglio Artistry in Landscaping" awards in the 2005 Arizona Department of Water Resources/Tohono Chul Park Xeriscape Contest.

We are increasing our push to improve the efficiency and convenience of our system to make rainwater the primary water source within our home as well. We *continually reasses what we have done*, and have expanded or modified basins where needed, replaced or relocated poorly selected or planted vegetation, and made our cistern and greywater systems more accessible and convenient to use. We are also planning an outdoor kitchen area with a new covered porch and cistern so we can experiment with rainwater for drinking, cooking, cleaning, and more.

Other people are developing their systems and experience in much the same way. We help, teach, and encourage each other, and the movement grows as the practice, examples, and knowledge within our backyard microwatersheds begin to overflow and nourish the greater community watershed. Start at the top. Start small. Start.

Appendix 1
Patterns of Water Flow and Erosion with Their Potential Water-Harvesting Response

PATTERNS OR TRACKS

The patterns or "tracks" left by the flow of water and sediment are excellent guides directing the selection and placement of water-harvesting efforts. Erosion is one such pattern. Erosion is a natural occurrence, which in healthy watersheds is naturally checked and slowed by vegetation and porous, living soils. In healthy watersheds erosion is slow and is a normal part of a dynamic equilibrium that moves sediments downstream. As organic matter and soil migrate downslope, they are replaced by on-site leaf drop from vegetation, soil migrating downwards from farther upslope, and the addition of digested plant material that "migrates uphill" and gets deposited in the form of animal droppings. All along the slope (except at the peak) the slow, downward migration of organic matter and soil is checked by its replacement.

In unhealthy watersheds, unchecked erosion can be like a deep cut in the human body, leading to a rapid loss of water and soil. It is a sign of depleting resources in an unstable landscape. Learning to recognize erosion patterns/tracks and their causes is an essential step in planning effective water-harvesting strategies that break the erosion triangle in figure A1.1.

SPEED, DISTANCE, and VOLUME refer to characteristics of water flowing on the land's surface. Reduce any of the three and you begin to cut the erosion cycle. The more you break the cycle the more you reduce erosion. If you put in a water-harvesting strat-

egy such as a berm 'n basin, check dam, or infiltration basin in the path of water flowing over the land, you will reduce erosion by reducing the SPEED of water flow. You'll be putting an earthen speed bump on the erosion highway.

If you place one or more of these strategies at the top of the watershed rather than beginning at the bottom, you will be reducing erosion by reducing the DISTANCE the water travels before being encouraged to infiltrate into the soil. This is like placing the speed bump at the top of the driveway so the water doesn't ever get a chance to destructively speed up.

If numerous strategies that hold and infiltrate runoff water into the soil are placed throughout a watershed—from tops of slopes to their bottom, erosion is reduced further by reducing the VOLUME of surface water flow. Surface water will not be able to accumulate into a destructive volume before being absorbed into the soil. Flowing overland in an erosive

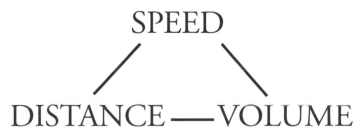

Fig. A1.1. Erosion triangle

manner will become so difficult for the water, it will choose to stay home in the soil or calmly walk through the landscape.

Below are a number of water and erosion flow patterns, and their potential response of water harvesting earthworks. The response techniques are covered in detail in volume 2 on earthworks.

PATTERN AND RESPONSE

SHEET FLOW

Sheet flow is the relatively even distribution of runoff water over the land surface, following the slope of the land downward, but not focused into distinct channels. Sheet flow has most likely occurred after a large rainfall if you don't see distinct channels in an area of sloping bare dirt. If the water hasn't focused into a channel, it must be crossing the land as sheet flow. Other indicators are microdetritus berms and plant pedestals described below.

Microdetritus Berms

PATTERN: Microdetritus berms are small, curved lines of organic matter such as leaf duff that has been carried by sheet flow, then settled out perpendicular to the flow. The outer bow of the curve usually points in the downward direction of the slope. They are typically less than two inches high, and often don't last more than a few weeks after a rain. They're found only on gentle slopes in yards or in the broad landscape, not in drainages. (See figure A1.2.)

RESPONSE: These tiny berms of organic matter are an indicator of calmer sheet flow, and therefore typically don't signal a pressing need for erosion control. They do, however, help confirm the direction of a gradual slope and water flow. This is useful when laying out contour berm 'n basins, boomerang berms, or planting on-contour.

Pedestals

PATTERN: Pedestals are mounds of soil held in place by grasses, shrubs, and low trees (fig. A1.3A). The

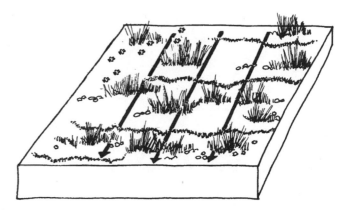

Fig. A1.2. Microdetritus berms and sheet flow

ground outside the perimeter of the pedestals is lower. A close look reveals that plant roots and protective canopies of leaves keep the pedestal from washing away. This same net-like canopy of leaves and branches also helps build the mounds by catching wind- and water-borne soil and organic matter, and by contributing leaf and twig drop to the soil below. The presence of pedestals usually indicates that more substantial sheet erosion is occurring within the broad landscape, though sometimes pedestals are observed in drainages where concentrated flow washes away sediments not held in place.

RESPONSE: Strategies that slow runoff and increase infiltration in the broad landscape, such as berm 'n basins, infiltration basins, imprinting, and increased vegetation, are usually appropriate to reduce substantial sheet erosion (fig. A1.3B).

CHANNEL FLOW

Channel flow is the concentrated distribution of runoff within distinct channels or drainages. Look for nick points, rills, gullies, bank cutting, different sediment sizes, vegetation growing within channels, and exposed roots. These patterns are described below.

Nick Point or Headcut

PATTERN: A nick point or headcut is an erosion feature created when sheet flow has concentrated into channel flow by cutting a nick or gouge into

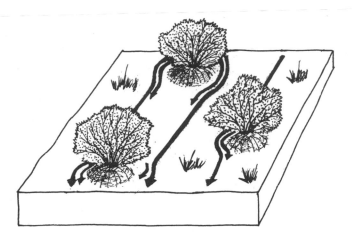

Fig. A1.3A. Pedestals and sheet flow

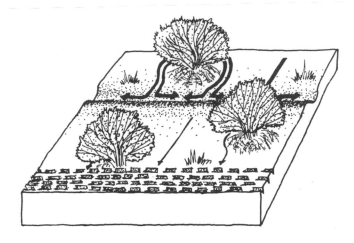

Fig. A1.3B. With added contour berm and imprints

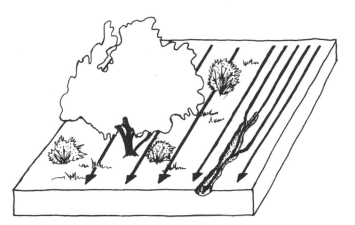

Fig. A1.4A. Nick point at head of rill

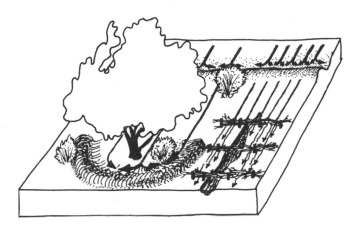

Fig. A1.4B. With added contour berm upslope, boomerang berm with tree, and brush weirs

the earth (fig. A1.4A). Nick points start small, but can grow to be quite severe. They are the growing edges of both rills and gullies, and will continue growing upstream, as long as there is soil to cut, or until they are corrected and stabilized. Thus the erosion moves upstream as water moves downstream.

RESPONSE: Nick points need to be quickly addressed to check erosion. Overland sheet flow should be slowed, spread out, and infiltrated into the soil as much as possible before it reaches a nick point. Water-harvesting strategies such as berm 'n basins, imprinting, vegetation, mulch, and infiltration basins may all be useful upslope of a headcut. Within the eroding channel itself, spread and infiltrate the flow

with permeable barriers appropriate to the scale of the channel (see below: "Rill Erosion" and "Gullies" for examples). See figure A1.4B.

Rill Erosion or Runnels

PATTERN: Rills or runnels are tiny erosive drainages in which loose soil has washed away. They are very common on eroding slopes where roadways have been cut into hillsides or on bare dirt driveways and roads that run downslope (fig. A1.5A). Rills are an early stage of the type of channel erosion that occurs downhill from nick points.

RESPONSE: Look first to spread and infiltrate sheet flow above the rill or runnel. Berm 'n basins,

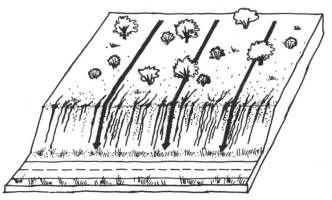

Fig. A1.5A. Rills on roadcut

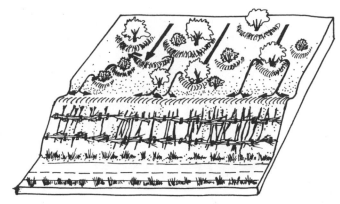

Fig. A1.5B. With added boomerang berms, contour berms, and brush weirs

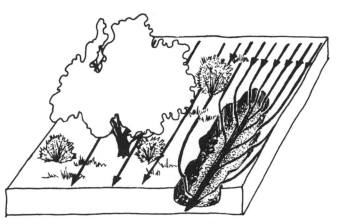

Fig. A1.6A. Gully

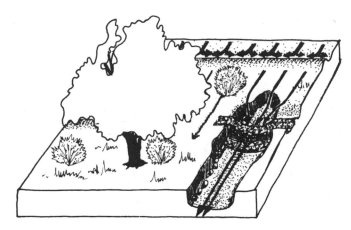

Fig. A1.6B. With added contour berm and gabion

vegetation, and mulch can all be effective. Then spread and infiltrate the flow within the rill itself, with a series of very small check dams constructed of branches and rock piles laid across the cut (fig. A1.5B).

Gullies

PATTERN: Gullies are large erosive drainages or arroyos. Oftentimes they were runnels or rills that continued to erode, deepen, and grow. Gullies are a channel erosion feature (fig. A1.6A).

RESPONSE: The overland flow draining toward the channel should be slowed, spread, and infiltrated into the soil as much as possible before reaching the drainage. Berm 'n basins, imprinting, vegetation,

mulch, and infiltration basins may all be appropriate. Within the drainage itself a series of sturdy, well-placed check dams constructed perpendicular to the flow can help stabilize the drainage as the broader landscape is simultaneously repaired (fig. A1.6B).

Bank Cutting at Curves

PATTERN: Bank cutting occurs where channelized water flows around a curve and the forward momentum of the water cuts the outside bank of the drainage. The slower moving water on the inside of the curve often allows this cut sediment to deposit at the inside curve location. Notice the shape of the cutting side of the flow, usually a vertical cliff. Then

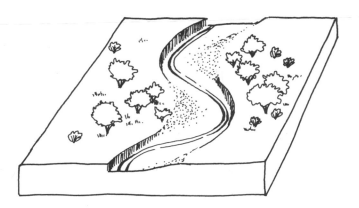

Fig. A1.7. Bank cutting at curves

Fig. A1.8. Different sizes of sediment

notice the shape of the deposition side, usually a gently sloping bank. (See figure A1.7.)

RESPONSE: Such bank cutting on a curve can help "spread and infiltrate" water by eventually widening and elongating the flow path via an ever more serpentine channel. This can reduce the gradient (slope) of a drainage and slow the water's flow. Do not attempt to control this curving of a channel. Rather use your recognition of the curving to correctly place check dams if erosion control within the channel is necessary. Check dams are water-harvesting and erosion-control structures placed perpendicular to the channelized water flow in straight sections of channels upstream from curves. Proper construction and placement of these structures helps ensure that bank cutting does not occur around the edges of the check dams.

Sediment Size within the Streambed

PATTERN: Different sizes of sediment are clues to past flows of water and sediment in a watercourse. The larger the sediment pieces, such as large rocks or boulders, the faster and stronger the past water flow was, and the potential flow could be. The presence of large rocks or boulders indicates fast-moving water, since these are the only objects heavy enough to settle out of fast-moving water. As the water slows, smaller and lighter sediment falls out. Stones are deposited first, then sand, and finally silts and clays when the water slows greatly. (See figure A1.8.) A boulder-strewn drainage may be dry at the time of observation, but when it floods it will flood with force.

RESPONSE: Determining potential flow is key to selecting and placing appropriate water-harvesting and erosion-control structures. It is usually better to stay out of boulder-strewn drainages with potentially intense flows. Instead, focus water-harvesting and erosion-control efforts on the more gentle flows of the broad landscape and on smaller drainages that feed the more intense flows.

Vegetation in the Bottoms of Drainages

PATTERN: Vegetation (or the lack of it) growing in the bottom of an arroyo gives you an idea of past flows. Notice the amount of down-cutting, or erosive deepening that has occurred in the drainage. Then try to determine the ages of the younger trees, perennial grasses, and other vegetation growing in the bed of the waterway.

RESPONSE: The size, density, and age of vegetation growing in a drainage bed and along its lower banks is a good indicator of flood frequency, as floods will usually scour out small vegetation. If you are trying to revegetate a drainage, the presence or lack of vegetation can "teach" you where it is acceptable to plant to reduce the risk of losing the new vegetation to flooding.

Exposed Roots

PATTERN: Exposed roots of trees and shrubs may be evident along large and small eroding drainages (fig. A1.9A).

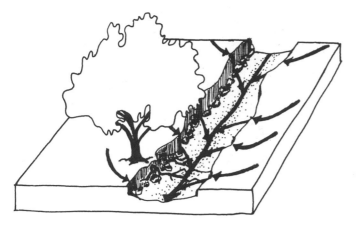

Fig. A1.9A. Exposed roots

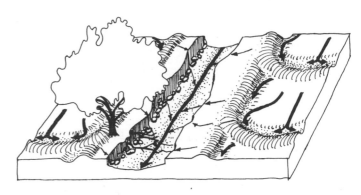

Fig. A1.9B. With added berm 'n basins

RESPONSE: A result of channel and bank erosion, exposed roots clue you in to the degree of erosion and potential water flow in a drainage. Generally, the more roots are exposed, the more intense the flow and the greater the depth and width of erosive cutting. Appropriate erosion-control strategies are generally the same as for gullies, though if the active bank cutting is too severe, first concentrate on reducing the severity of the flow by harvesting water and reducing erosion higher in the watershed. (See figure A1.9B.)

GENERAL PATTERNS OF WATER, SLOPE, AND FLOW

The following patterns are not limited to sheet or channel flow alone. They are caused by various flows of water, the life forms it supports, and the slopes it helps shape. Look for sediment deposition, break lines and keylines, high water marks, scour holes, vegetation, and animals. These patterns are described below.

Sediment Deposition

PATTERN: Sediment deposition occurs where rocks, sand, soil, twigs, seeds, animal droppings, and other materials carried by runoff have dropped out of the flow. Branches, stones, and patches of grass or shrubs occasionally capture these sediments. Look for these deposits on the broad landscape where sheet flow is present or in drainages, and speculate about their source.

RESPONSE: The presence of different types of sediments can give you a good idea of the size and extent of your watershed. If you're in a low desert valley and oak, walnut, or pine debris is present, your watershed probably extends into the higher elevations where that vegetation naturally occurs. If grass clippings and mulberry leaves appear in your yard, but you have neither a lawn nor a mulberry tree, search for their upslope source.

Break Lines and Keylines

PATTERN: Break lines are places in the landscape where slopes change from gentle grades where sediments settle out of slow moving runoff, to steep grades where sediments are picked up and carried away by faster moving runoff. Keylines are where slopes change from steep grades where sediments are picked up and carried away by fast moving runoff, to more gentle grades where sediments settle out of slower moving runoff. On a micro-level, you may see a break line where a slope covered with leaves and silts changes to a steeper patch of naked sloping dirt. The keyline would be downhill where that naked slope changes to a more gradual slope where collected fines, silts, and organic matter accumulate. The land below that keyline could be a good location for planting, since it is not eroding, but instead is receiving water and organic matter. On a macro-level, a mountain or hilltop slopes down to a breakline where it steepens and erosion increases, then comes the keyline where

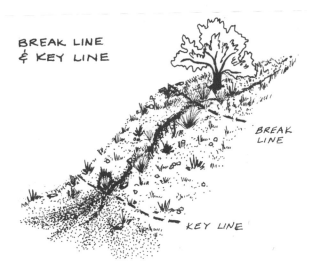

BREAK LINE
& KEY LINE

BREAK
LINE

KEY LINE

Fig. A1.10. Break line and keyline

Fig. A1.11. High water flow detritus
on young cottonwood

the slope lessens at the top of an alluvial fan composed of depositing sediments. (See figure A1.10.)

RESPONSE: Concentrate water-harvesting efforts where less effort is required—above break lines first, then below keylines second. Be wary of the slope between the break line and the keyline as it may be too steep or challenging to do more than plant vegetation on-contour, or carefully lay out contour berms or terraces made of brush or single courses of rock. If you are building a small earthen dam or pond, make sure you will not back water up above a keyline. You need to keep the water level below the keyline, so you can direct your overflow across more gentle, easily managed slopes.

High Water Marks

PATTERN: High water marks are the highest points in drainages or floodplains where you can see evidence of past high water flow. Look for lines of discoloration on rocks and vegetation and deposits of branches, twigs, grass, and other debris that indicate the high water mark from flooding. Sometimes these deposits are surprisingly high on a fence or in the branches of a tree. Within your property or yard make sure no high water marks appear above or near the level of your home's foundation. (See figure A.11.)

RESPONSE: High water marks tell you of potential high-water events in drainages and floodplains. No strategy is used to change this on the broad landscape; just don't build, or plant extensively, within the area of potential flooding. If you find evidence of water backing up onto a building's foundation, try to harvest that water with earthworks before it reaches the building, and make sure the grade around the building drains water to a point 10 feet (3 m) away from the building. Urban flood peaks can become more intense as more land in the watershed gets covered in roofs, streets, and parking lots. It is not wise to buy a home located on flood-prone land.

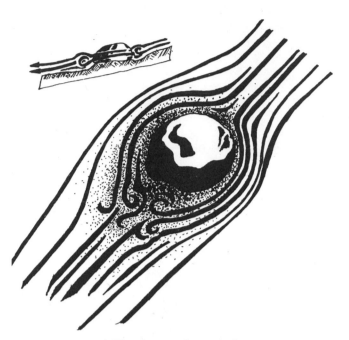

Fig. A1.12. Scour hole

Scour Holes

PATTERN: Scour holes form in locations where water is forced around an immovable object such as a boulder, often creating a whirlpool directly downstream of the obstruction. These whirlpools leave behind distinctive circular holes in the bottoms of arroyos, and on broad landscapes experiencing sheet flow (fig. A1.12).

RESPONSE: Scour holes do not necessarily indicate a need for erosion control, but can help confirm the direction and force of potential water flow even when no water is currently flowing.

Vegetation

PATTERN: Vegetation is generally densest where water is present. Hydric vegetation—plants that tolerate some waterlogging—flag water at or close to the surface. In the Southwest, water-needy broad-leafed cottonwood (*Populus fremontii*) and sycamore trees (*Platanus wrightii*) typically indicate the presence of springs, perennial water flow, or shallow groundwater levels. Hardy triangle-leaf bursage shrubs (*Ambrosia deltoidea*) are found in arid, drained zones. Everything is relative; however, in extremely dry areas sometimes even bursage cannot grow in the arid drained zones of the broad landscape, and instead is found along drainages and other areas of greater water concentration. Short-term indicators of soil moisture include native annuals such as peppergrass (*Lepidium thurberi*) and abundant invasive annuals such as tumbleweed (*Salsola iberica*).

RESPONSE: Familiarize yourself with the water needs of local plants. They'll tell you how much water is in the soil and what types of vegetation with similar water needs the landscape can support.

Animals

PATTERN: Animals, insects, and birds that need water can signify the proximity or dependability of a water source. Dragonflies are found near open bodies of water. A high number of toads probably mean that a water source is ephemeral or too small to support predatory fish.

RESPONSE: Familiarize yourself with the water needs of local animals, insects, and birds. When assessing a site, use sightings or evidence of these life forms to clue you in to local water sources. In urban settings with fountains and pools, the presence of water-dependent creatures may not indicate the presence of natural water supplies.

Appendix 2

Water Harvesting Traditions in the Desert Southwest

By Joel Glanzberg
Illustrated by Roxanne Swentzell

This article first appeared in the August 1994 Permaculture Drylands Journal
and is reprinted with the permission of the Permaculture Drylands Institute.

Before the advent of modern irrigation technology, peoples of the American Southwest relied upon an array of water-harvesting and water-conserving techniques to grow their food. Not only are these techniques still appropriate, but their use, scale, and at times, failures have much to teach us. Several of the systems used by traditional peoples are described and illustrated below.

CHECK DAMS

Check dams are built across drainages that flow only periodically. They are constructed of rock and can range in size from small to large. These rock dams catch soil and water, and were often built higher as more soil accumulated behind them. They provided an excellent way to fertilize soil and stabilize drainages,

Fig. A2.1. Check dams

and were used for all kinds of crops. There are good examples in Colorado, Utah, and in New Mexico—at Mesa Verde and Hovenweep National Parks, at numerous small dams in the upper Rio Grande and Chama drainages, and throughout the Pajarito Plateau (fig. A2.1).

TERRACES OR LINEAR BORDERS

Terraces themselves were occasionally built, but linear borders or low lines of stone across slopes of hills were more common. At Point of the Pines in Arizona, hilltop pueblos were surrounded by concentric rings of rocks gathered from the entire hillside and laid along contour lines across the slopes. Soil washing down the bare hillsides caught behind the stonewalls, accumulating up to 16" deep. This loose soil would have been highly fertile and water absorbing (fig. A2.2).

WAFFLE GARDENS

Waffle gardens can be either sunken beds with ground-level berms, or ground-level beds surrounded by raised berms of earth. The bermed beds catch and hold rainwater as well as retain water brought by hand. Waffle beds were built on a very small scale for especially valuable crops. They are best known historically at Zuni (fig. A2.3).

Fig. A2.2. Linear borders

Fig. A2.4. Grid gardens

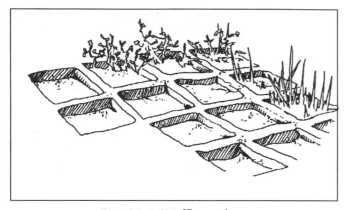

Fig. A2.3. Waffle gardens

GRID GARDENS

Grid gardens are similar to waffle gardens, but have walls made of stone rather than earth. They usually have much larger beds than waffle gardens. The walled beds help hold soil and were often placed to catch water runoff from moderate slopes. They were used extensively during prehistoric times throughout the upper Rio Grande, the Chama, and the Ojo Caliente drainages. They were probably not hand-watered, and it appears likely that they were used to grow major crops such as corn and beans (fig. A2.4).

GRAVEL- AND ROCK-MULCHED FIELDS

Mulch of any kind slows evaporation by sheltering the soil surface. The Anasazi clearly knew this. Throughout the upper Rio Grande and Chama Drainages, vast areas were mulched with gravel. Grid gardens were often covered with mulch. Gravel mulches not only conserve moisture, they also reduce wind and water erosion. Dark gravel mulches increase soil and air temperatures, reducing the threat of early and late frosts. At Wupatki in northern Arizona, the Sinagua people were able to grow food without supplemental water largely because of the natural covering of cinders created by the eruption of Sunset Crater. This eruption, and the resultant layer of cinders that covered highly fertile, water-holding volcanic ash, was responsible for Anasazi, Hohokam, and Mogollon people moving into this area to live, creating the Sinagua culture around A.D. 1000 (fig. A2.5).

CLIFFBASE PLANTINGS

Often the water-collecting surfaces of cliffs were used to provide water for crops. By planting where water would run off and be concentrated, available moisture and fertility could be increased. At

Fig. A2.5. Gravel and rock-mulched fields

Fig. A2.6. Cliffbase plantings

Chaco Canyon, this technique was used extensively. A complex irrigation system was developed using cliff runoff. Grid gardens, check dams, and terraces were located in various places to catch this runoff (fig. A2.6).

FLOODPLAIN FARMING

Soil located in or near a channel of flowing water is usually moist and fertile. For this reason, floodplain fields were situated along the margins of permanent or ephemeral streams, the low terraces of arroyos, or within the bottoms of arroyos (fig. A2.7). The principle is similar to the moisture and fertility enhancement utilized by check dam agriculture located in steeper arroyos. In this case, flatter areas in drainages were planted, where a raised water table was also useful to the cultivator.

One disadvantage to such sites is cold air drainage into these canyon bottoms. The accumulation of cold air makes these sites susceptible to late spring and early fall frosts, limiting the length of the growing season. Other disadvantages limiting use of these areas are the danger of floods wiping out fields and the difficulty of clearing thick riparian growth.

Often brush weirs or earthen walls were used to slow or spread water across the fields. This led to irrigation, which spread the water over more land, enabling more crops to be grown with more control.

Fig. A2.7. Floodplain farming

Fig. A2.8. Floodwater farming

FLOODWATER FARMING

Often the fans of soil below arroyos or small canyons were planted to utilize the flood waters coming down these drainages. This is sometimes known as ak-chin farming and is still practiced by the O'odham and Hopi. In some cases, the field is sited beside the path of the arroyo. Brush weirs are then constructed across the bed of the arroyo to direct water out of the arroyo and onto the field. A destructive flood will blow out the weir but will not destroy the field. Destruction of the field by a flood is a very real problem with ak-chin farming for fields located directly in the paths of arroyos (fig. A2.8).

IRRIGATION

That there was irrigation before the arrival of the Spanish is clear. Its exact extent and character is not. What we do know is that it was not as universal a trait of Pueblo agriculture in the past as it is now, and was only one of a wide range of farming techniques used.

It is not clear to what extent the Rio Grande and other large rivers in northern New Mexico were used prehistorically for irrigation. Modern farming eliminates the evidence of previous irrigation systems, and scholars do not agree on the documentary evidence. At Santa Clara Pueblo, the looser soils in the canyon and immediately surrounding the village seem to have been preferred sites for irrigation. These areas were watered by Santa Clara Creek.

As late as 1940, almost all irrigated fields were located in the vicinity of Santa Clara Creek. There

Fig. A2.9. Irrigation

appears to have been only a gradual shift of landholding from the area around the pueblo to the river bottom, indicating that at Santa Clara, river bottom cultivation was one small part of Pueblo agriculture. Typically, fields were scattered across the landscape at different elevations and in different environments to prevent a disaster from destroying all of the crops. The Hohokam of southern Arizona built substantial irrigation canals to divert water from large rivers, but the eventual failure of these projects contributed to the destruction of their "advanced" civilization. In the Anasazi area, dry farming seems to have been the rule, and irrigation was small in scale compared to the Hohokam (fig. 2.9).

FOLLOWING THE ANASAZI THROUGH TIME

The appearance of the various farming techniques through time gives us a view into the environmental effects of Anasazi agriculture and their responses to these effects. Originally simple swidden agriculture was practiced. Land was cleared, burned, and planted. As it was exhausted, new land was cleared. Eventually the original plot recovered and could be replanted. This extensive clearing increased erosion. Check dams and linear borders were being constructed late in the occupations of Chaco, the Mimbres area, the San Juan Basin, and other sites, such as Pot Creek Pueblo near Taos. These structures were apparently an attempt to halt the serious erosion caused by deforestation and clearing, overuse of wild plants, and foot traffic. Despite these conservation attempts, these areas were ultimately abandoned. When a prolonged drought struck in A.D. 1276–1299, the food-production systems were already under stress from the high population densities. The combination of the drought with the environmental degradation caused by heavy farming and residential use probably led to final abandonment of settlements.

Refugees from these areas built grid gardens, techniques intended to prevent the start of erosion. Rather than waiting for erosion to begin, farmers were now attempting to stop the process at its source before it began, an example of Anasazi farmers learning from past mistakes and adapting to their environments.

As noted above, at an earlier time, the Anasazi had rotated fields, farming one until it was exhausted, and then clearing another, and so on until they returned to the first, many years later. As populations increased, however, the inhabitants were forced to use all available farmland in proximity to their villages. This forced the Anasazi to be somewhat nomadic, moving every 60–100 years when they had depleted a site's soil and other natural resources. After a period of time, the original group or another group could reinhabit an area, the soil fertility and natural resources having recovered from previous usage. This is a key part of pre-Colombian, Pueblo land use patterns. Even after the adoption of corn, squash, and beans led them to a sedentary lifestyle in villages, they continued to be semi-nomadic at a much slower pace.

It was continuous habitation and the associated large-scale populations, irrigation systems, building projects, deforestation, and soil depletion that contributed to the forced migrations of the 1300s. The land taught the Anasazi to keep things small and to move occasionally to allow it to rest. By remaining fluid within their environment, by using many techniques in various locations, microclimates, and elevations, and by maintaining an appropriately small scale, the Anasazi were able to survive where earlier growth and urbanization had failed. It is ironic that rather than learning from what has failed before and adopting what has succeeded, we have done the opposite. Like the Chacoans and Hohokam, we believe that our technical "advances," power, and grandeur make us exceptions to the constraints of our environment. And just like them, our failure will come to us as a surprise.

Joel Glanzberg is a master drylands permaculture teacher and designer living in northern New Mexico and working with the Regenesis Group (www.regenesis-group.com).

Roxanne Swentzel is an exceptional artist and co-creator (with Joel) of the thriving Flowering Tree Permaculture Site in the Santa Clara Pueblo, New Mexico.

Appendix 3
Water-Harvesting Calculations

List of Equations and Other Information

Box A3.1. Abbreviations, Conversions, and Constants for English and Metric Measurement Units

Note: * items are approximate or rounded off

ABBREVIATIONS FOR ENGLISH UNITS
inches = in
feet = ft
square feet = ft^2
cubic feet = ft^3
gallons = gal
pounds = lb
pounds per square inch of pressure = psi

CONVERSIONS FOR ENGLISH UNITS
To convert cubic feet to gallons, multiply cubic feet by 7.48 gal/ft^3 *
To convert inches to feet, divide inches by 12 in/ft
To convert gallons of water to pounds of water, multiply gallons by 8.34 lb/gal *
To convert cubic feet of water to pounds, multiply cubic feet by 62.43 lb/ft^3 *

CONSTANTS
Pounds of pressure per square inch of water per foot of height = 0.43 psi/ft *
Ratio between a circle's diameter and its circumference is expressed as π = 3.14 *

ABBREVIATIONS FOR METRIC UNITS
millimeters = mm
centimeters = cm
meters = m
liters = l
kilograms = kg

CONVERSIONS FOR METRIC UNITS
1 liter of water weighs 1 kilogram
To convert cubic centimeters to liters, divide cubic centimeters by 1,000

CONVERTING BETWEEN ENGLISH UNITS AND METRIC UNITS
To convert inches to millimeters, multiply inches by 25.4 mm/in *
To convert inches to centimeters, multiply inches by 2.54 cm/in *
To convert feet to meters, multiply feet by 0.30 m/ft *
To convert gallons to liters, multiply gallons by 3.79 liter/gal *
To convert pounds to kilograms, multiply pounds by 0.45 kg/lb *

Best technique to measure rainfall: Buy a simple rain gauge for $10 or so from a hardware or feed store, plant and garden nursery, or a scientific supply house. A rain gauge that is tapered at the bottom makes reading small amounts of rainfall easier.

For resources documenting local rainfall rates and other climatic information, see appendix 6, section G.

Equation 1A.
Catchment Area of Rectangular Surface (English units)

length (ft) × width (ft) = catchment area (ft²)

EXAMPLE:

A house that measures 47 feet long by 27 feet wide at the drip line of the roof. Note that it does not matter whether the roof is flat or peaked; the roof dimensions at the drip line are the same. It is the "footprint" of the roof's drip line that matters.

47 ft × 27 ft = 1,269 ft²
1,269 ft² = catchment area

If the roof consists of two or more rectangles, calculate the area for each rectangle and add together. Again, take the view of a falling raindrop, and only look at the "footprint" of the roof's drip line. Roof pitch cannot be seen from above and does not matter. With conical, octagonal, or other non-standard roof shapes, again calculate the area based on the *drip line*.

Equation 1B.
Catchment Area of Rectangular Surface (metric units)

length (m) × width (m) = catchment area (m²)

EXAMPLE:

15 m × 9 m = 135 m²
135 m² = catchment area

Again, all the considerations in Equation 1A will apply.

Equation 2A.
Catchment Area of Triangular Surface (right triangle)

Multiply the lengths of the two shorter sides of the triangle then divide by 2 = catchment area

EXAMPLE:

A triangular section of roof measures 9 feet by 12 feet by 15 feet. This is a right triangle, with the 90-degree angle between the 9-foot and 12-foot sides. Taking the measurements of the two shorter sides:

(9 ft × 12 ft) ÷ 2 = catchment area (ft²)
108 ft² ÷ 2 = 54 ft²

54 ft² = catchment area

Equation 2B.
Catchment Area of Triangular Surface (standard math formula)

Multiply the triangle's base times its height then divide by 2 = catchment area
where the base can be any side, and the height is measured perpendicularly from the base to the opposite vertex.

EXAMPLE:

You want to know the area of a triangular section of patio. The length of the section in front of you is 20 feet (triangle base) and you measure 4 feet perpendicularly to the opposite vertex of the triangle.

(20 ft × 4 ft) ÷ 2 = catchment area (ft²)
80 ft² ÷ 2 = 40 ft²

40 ft² = catchment area

Equation 2C.
Catchment Area of Triangular Surface (Heron's formula)

This formula, attributed to Heron of Alexandria (first century A.D.), involves no trigonometry. It only needs the square root (sqrt) function found on most electronic or computer calculators. It may be useful when dealing with non-right triangles where you can measure (or know) all sides of the triangle.

Step 1: Determine the lengths of the sides of the triangle. These are a, b, c.

Step 2: Calculate s.
(a + b + c) ÷ 2 = s

Step 3: Calculate S, using:
s × (s − a) × (s − b) × (s − c) = S

Step 4: Calculate the catchment area, which is the square root of S.
sqrt S = catchment area

Equation 3.
Catchment Area of Circular Surface

$\pi \times r^2$ = catchment area
Note: r = radius of the circle. A circle's radius is half the circle's diameter.

EXAMPLE:

A circular roof has a 25 foot diameter. Divide the diameter by 2 to get the *radius* of 12.5 feet.

π × (12.5 ft × 12.5 ft) = catchment area (ft²)
3.14 × 156.25 ft² = 490.62 ft²

490.62 ft² = catchment area

Equation 4A.
Possible Volume of Runoff from a Roof
or Other Impervious Catchment Area (English units)

catchment area (ft²) × rainfall (ft) × 7.48 gal/ ft³ = maximum runoff (gal)

Note: For a more realistic and conservative estimate see Equation 5.

EXAMPLE CALCULATING ANNUAL RUNOFF:

Calculate the gallons of rain running off the roof in an average year from a home that measures 47 feet long and 27 feet wide at the drip line of the roof. (In the example below, the roof dimensions at the drip line are included in the calculation; the catchment area is the same whether the roof is flat or peaked.) Rainfall in this location averages 10.5 inches per year, so you will divide this by 12 inches of rainfall per foot to convert inches to feet for use in the equation. (Note: You can use the same equation to calculate the runoff from a single storm, by simply using the rainfall from that storm instead of annual average rainfall in the equation.) Since the roof is a rectangular area, use the following calculation for catchment area:

(length (ft) × width (ft)) × rainfall (ft) × 7.48 gal/ft³ = maximum runoff (gal)
(47 ft × 27 ft) × (10.5 in ÷ 12 in/ft) × 7.48 gal/ft³ = maximum runoff (gal)
1,269 ft² × 0.875 ft × 7.48 gal/ft³ = 8,306 gal

8,306 gal = runoff

EXAMPLE CALCULATING RUNOFF FROM A SINGLE RAIN EVENT:

Calculate the maximum gallons of rain running off the roof in a single rain event from a home that measures 47 feet long and 27 feet wide at the drip line of the roof. It is not unusual for heavy storms in the example area to drop two inches of rain. To determine the runoff from such a rain event you will divide the 2 inches of rainfall by 12 inches of rainfall per foot to convert inches to feet for use in the equation. Since the roof is a rectangular area, use the following calculation for catchment area:

(length (ft) × width (ft)) × rainfall (ft) × 7.48 gal/ft³ = maximum runoff (gal)
(47 ft × 27 ft) × (2 in ÷ 12 in/ft) × 7.48 gal/ft³ = maximum runoff (gal)
1,269 ft² × 0.167 ft × 7.48 gal/ft³ = 1,585 gal

1,585 gal = maximum runoff

Equation 4B.
Possible Volume of Runoff from a Roof or
Other Impervious Catchment Area (metric units)

catchment area (m²) × rainfall (mm²) = maximum runoff (liters)

Calculations for annual rainfall, a rainy season, or an event would be similar to those for English units.

Box A3.2. Estimating Rainfall Runoff Using Rules of Thumb

Rough rule of thumb for calculating rainfall runoff volume on a catchment surface (English units):
You can collect 600 gallons of water per inch of rain falling on 1,000 square feet of catchment surface.

On the really big scale:
You can collect 27,000 gallons of water per inch of rain falling on 1 acre of catchment surface.

Rule of thumb for calculating rainfall volume on a catchment surface (metric units):
You can collect 1,000 liters of water per each 10 millimeters of rain falling on 100 square meters of catchment surface.

On the really big scale:
You can collect 100,000 liters of water per 10 millimeters of rain falling on one hectacre of catchment surface.

Equation 5A.
Estimated Net Runoff from a Catchment Surface Adjusted by its Runoff Coefficient (English units)

catchment area (ft^2) $\times$ rainfall (ft) $\times$ 7.48 gal/ft $\times$ runoff coefficient = net runoff (gal)

Impervious catchment surfaces such as roofs or non-porous pavement can lose 5% to 20% of the rain falling on them due to evaporation, and minor infiltration into the catchment surface itself. The more porous or rough your roof surface, the more likely it will retain or absorb rainwater. On average, pitched metal roofs lose 5% of rainfall, allowing 95% to flow to the cistern. Concrete or asphalt roofs retain around 10%, while builtup tar and gravel roofs can retain 15% to 20%. However, the percent of retention is a function of the size and intensity of the rain event, so more porous roof surfaces could absorb up to 100% of small, light rain events. To account for this potential loss, determine the runoff coefficient that is appropriate for your area and impervious catchment surface (0.80 to 0.95).

EXAMPLE CALCULATING NET ANNUAL RUNOFF FROM A ROOF:

Calculate the net gallons of rain running off the roof in an average year from a home that measures 47 feet long and 27 feet wide at the drip line of the roof. Rainfall in this location averages 10.5 inches per year, so you will divide this by 12 inches of rainfall per foot to convert inches to feet for use in the equation. (Note: You can use the same equation to calculate the runoff from a single storm, by simply using the rainfall from that storm instead of annual average rainfall in the equation.) Assume that the loss of water that occurs on the catchment surface is at the high end of the range so you get a conservative estimate of *net runoff*. This means you select a *runoff coefficient* of 80%, or 0.80. Since the roof is a rectangular area, use the following calculation for catchment area:

(length (ft) $\times$ width (ft)) $\times$ rainfall (ft) $\times$ 7.48 gal/ft^3 $\times$ 0.80 = net runoff (gal)
(47 ft $\times$ 27 ft) $\times$ (10.5 in $\div$ 12 in/ft) $\times$ 7.48 gal/ft^3 $\times$ 0.80 = net runoff (gal)
1,269 ft^2 $\times$ 0.875 ft $\times$ 7.48 gal/ft^3 $\times$ 0.80 = 6,644 gal

6,644 gal = net runoff

Based on this, a realistic estimate of the volume of water that could be collected off the 47 foot by 27 foot example roof in an average year is 6,644 gallons.

Pervious surfaces such as earthen surfaces or vegetated landscapes can infiltrate up to 100% of the rain falling on them. Their runoff coefficient is greatly influenced by soil type and vegetation density. Large-grained porous sandy soils tend to have lower runoff coefficients while fine-grained clayey soils allow less water to infiltrate and therefore have higher runoff coefficients. Whatever the soil type, the more vegetation the lower the runoff coefficient since plants enable more water to infiltrate the soil. Below are some runoff coefficients for the southwestern U.S., although these are just rough estimates since runoff rates are also affected by rainfall intensity and duration. The more intense or the longer the rainfall the greater the runoff, since more rain is infiltrated in the soil before the soil becomes saturated. A very light rainfall may just evaporate, and not run off or infiltrate at all.

- Sonoran Desert uplands (healthy indigenous landscape): range 0.20–0.70, average 0.30–0.50
- Bare earth: range 0.20–0.75, average 0.35–0.55
- Grass/lawn: range 0.05–0.35, average 0.10–0.25
- For gravel use the coefficient of the ground below the gravel

EXAMPLE CALCULATING NET *ANNUAL* RUNOFF FROM A *BARE SECTION OF YARD*:

In an area receiving 18 inches of rain in an average year, you want to calculate the runoff from a 12 foot by 12 foot bare section of yard that drains to an adjoining infiltration basin. The soil is clayey and compacted, and you estimate its runoff coefficient to be 60% or 0.60.

catchment area (ft^2) × rainfall (ft) × 7.48 gal/ft × runoff coefficient = net runoff (gal)
12 ft × 12 ft × (18 in ÷ 12 in/ft) × 7.48 gal/ft^3 × 0.60 = net runoff (gal)
144 ft^2 × 1.5 ft × 7.48 gal/ft^3 × 0.60= 969 gal

969 gal = net runoff

Based on this, a realistic estimate of the volume of runoff that could be collected off the 12 foot by 12 foot section of bare earth within the adjoining infiltration basin is 969 gallons in an average year.

EXAMPLE CALCULATING RUNOFF FROM A *SINGLE STORM EVENT* ON *ESTABLISHED LAWN (GRASS)*:

The runoff coefficient for this established lawn is assumed to be 20% or 0.20, and the maximum storm event is 3 inches:

12 ft × 12 ft × (3 in ÷ 12 in/ft) × 7.48 gal/ft^3 × 0.20 = net runoff (gal)
144 ft^2 × 0.25 ft × 7.48 gal/ft^3 = 54 gal

54 gal = net runoff

Equation 5B.
Estimated Net Runoff from an Impervious Catchment Surface Adjusted by its Runoff Coefficient (metric units)

catchment area (m²) × rainfall (mm) × runoff coefficient = net runoff (liters)

EXAMPLE:

In an area receiving 304 millimeters of rain a year, you have a rooftop catchment surface that is 15 meters long and 9 meters wide, and you want to know how much rainfall can realistically be collected off that roof in an average year. You want a conservative estimate of annual *net runoff*, so you use a *runoff coefficient* of 80% or 0.80. (Since the roof is a rectangular area, use the following calculation for catchment area as in Equation 1B— catchment area (m²) = length (m) × width (m)—which is figured into the calculation below.)

(length (m) × width (m)) × rainfall (mm) × 0.80 = net runoff (liters)
(15 m × 9 m) × 304 mm × 0.80 = net runoff (liters)
135 m² × 304 mm × 0.80 = 32,832 liters

32,832 liters = net runoff

A realistic estimate of the volume of water that could be collected off this 15 meter by 9 meter roof in a year of average rainfall is 32,832 liters.

Equation 6.
Cistern Capacity Needed to Harvest the Roof Runoff from a Large Storm Event

catchment area (ft²) × rainfall expected in a local high volume storm (ft) × 7.48 gal/ft² × runoff coeficient = catchment runoff (gal)

EXAMPLE:

A water harvester with a 1,200 ft² roof lives in an area where a single storm (or two storms just a few days apart) can unleash 3 inches of rain.

1,200 ft² × (3 inches ÷ 12 inches) × 7.48 gal/ft² × 0.80 = catchment runoff (gal)
1,200 ft² × 0.25 ft × 7.48 gal/ft² × 0.80 = 1,795 gal

1,795 gal = catchment runoff

This is the *minimum cistern volume* needed to capture the roof runoff for this size storm.
Note: The above calculation is meant to give a rough estimate of a tank size that will reduce water loss to overflow from the tank and extend the availability of a lot of rainfall long after the rain event only—it is not based on estimated water needs. It is a quick and easy calculation for those simply wanting to supplement their water use with efficient rainwater tank storage. I often recommend beginner water harvesters start with a tank not exceeding a 1,500 gallon capacity. The system can always be expanded later. To start small you don't need to begin with a tank harvesting all the roof's runoff; rather begin by sizing a tank capturing water from just one section of the roof.

Equation 7.
Water Storage Capacity Needed for a Household Committing to Use Harvested Rainwater as the Primary Water Source (English units)

number of people × daily water consumption (gal/person/day) × longest drought period (days) = needed storage capacity (gal)

EXAMPLE:

If three people live in the household used in the above examples, each person consumes an average of about 50 gallons per day, and the typical dry season in their area lasts 140 days then:

3 people × 50 gal/person/day × 140 days = 21,000 gal

21,000 gal = needed water capacity

If the people in this household are planning to live primarily off rainwater at their current water consumption rate they would be wise to plan for at least 21,000 gallons of water collection and storage capacity to get them through up to 140 days of dry times.

If the needed water capacity (and needed catchment area) seems too large to be feasible, see how much you can realistically reduce your water consumption, then do the calculation again. For example, if the same household could reduce its daily water consumption to 20 gallons/person/day only 8,400 gallons of water collection and storage capacity would be needed.

Note: The above calculation will give a ballpark estimate of minimum tank capacity to meet dry season demand in expected drought. Sufficient catchment directing water to the tank is also needed to ensure the tank is full or close to full on day one of the dry season. See volume 3 of *Rainwater Harvesting for Drylands*, for additional calculations and considerations.

Equation 8. Potential Gravity-Fed Water Pressure from Your Tank (English units)

height of water above its destination (ft) × water pressure per foot of height (psi/ft) = passive water pressure (psi)

For every foot your source of water is above the elevation of the place where it will be used you develop 0.43 psi/ft of passive water pressure (gravity is the only force being used to create that pressure). The source of water may be in a tank, or a gutter and its associated downspout. The place you use the water may be a garden bed, a fruit tree basin, or any other location where supplemental water is needed.

EXAMPLE:

The folks with the new 8-foot-tall tank want to figure out how much passive water pressure will be available to deliver water from the tank to their squash plants placed in basins 6 inches (0.5 ft) below the surrounding land surface. The height of water in the 8-foot tank is 4 inches below the top of the tank due to the presence of an overflow pipe that allows excess rainwater to safely flow out of the tank during large storms. Based on this information the height of water above its destination is around 8.1 ft. Using Equation 8, calculate the passive water pressure as follows:

8.1 ft × 0.43 psi/ft = 3.48 psi

3.48 psi = passive water pressure

As the cistern water is used, the water pressure will drop with the dropping level of water (head) in the tank. Also, keep in mind that friction between water and the walls of a hose, pipe, or irrigation line will cut down on water pressure, so to maintain pressure try to use the water close to the tank, reducing the length of pipe or hose. For example, place a garden on the east side of your tank where the veggies will be shaded from the hot afternoon sun by the bulk of the tank, and you won't need a hose any longer than 25 feet (7.6m).

EXAMPLE:

I often place cisterns so their base is at least 2.5 feet above the garden or basin receiving the stored water. This guarantees me at least 1 psi of gravity-fed pressure even when the tank is nearly empty.

height of water above its destination (ft) × water pressure per foot of height (psi/ft) = passive water pressure (psi)
2.5 ft × 0.43 psi/ft = 1.08 psi

1.08 psi = passive water pressure

Equation 9A.
Storage Capacity of a Cylinder (Can Apply to Both a Cylindrical Cistern or a Length of First Flush Pipe) (English units)

π × (cylinder radius (ft))^2 × effective cylinder height* (ft) × 7.48 gal/ft^3 = capacity (gal)

Note: r = radius of the circle

*Effective height is the height of water you can get back out of the tank when it's full, as opposed to the total height of water in the tank, which includes several inches of water that can never be drained out due to an outflow pipe above the bottom of the tank.

EXAMPLE:

The householders above are considering using a cylindrical tank to store their rainwater. They want to determine the capacity of a tank with a diameter of 3 feet and a height of 8 feet. The radius of the tank is one half the diameter, so it is 1.5 feet. Since they realize the effective tank storage height is going to be reduced by 4 inches because of the raised outlet 4 inches from the bottom of the tank, and by another 5 inches because of the bottom of the tank overflow pipe being 5 inches below the top of the tank, the effective height is going to be 7.25 feet. Using Equation 9A, they calculate the usable capacity of the tank as follows:

π × (1.5 ft)^2 × 7.25 ft × 7.48 gal/ft^3 = capacity (gal)
3.14 × 2.25 ft^2 × 7.25 ft × 7.48 gal/ft^3 = 383 gal

383 gal = capacity

Equation 9B.
Storage Capacity of a Cylinder (Can Apply to Both a Cylindrical Cistern or a Length of First Flush Pipe) (metric units)

$\pi \times (r\ (cm))^2 \times$ effective cylinder height (cm) $\div$ 1,000 cm³/liter = capacity (liters)

See notes for Equation 9A.

Equation 10A.
Storage Capacity of a Square or Rectangular Tank (English units)

length (ft) × width (ft) × effective height (ft) × 7.48 gal/ft³ = capacity (gal)

EXAMPLE:

A household decides to install a rectangular tank that has interior dimensions: 8 feet tall, 6 feet long, and 4 feet wide. The tank outlet tap is located 4 inches above the bottom of the tank. The underside of the overflow pipe is located 5 inches below the top of the tank. They calculate the effective height of water as 7.25 ft, so the calculation is as follows:

6 ft × 4 ft × 7.25 ft × 7.48 gal/ ft³ = 1,301 gal

1,301 gal = capacity

Equation 10B.
Storage Capacity of a Square or Rectangular Tank (metric units)

length (cm) × width (cm) × effective height (cm) ÷ 1,000 cm³/liter = capacity (liters)

See notes with Equation 10A.

Equation 11A.
Cistern's One-Time Dollar Price for Storage Capacity (English units)

price of cistern (dollars) ÷ storage capacity (gal) = price of storage capacity (dollars/gal)

EXAMPLE:

The tank in Equation 9A holds 1,301 gallons of water, and would cost around $850 to purchase and install:

$850 ÷ 1,301 gal = $0.65/gal

$0.65/gal = price of storage capacity

Equation 11B.
Cistern's One-Time Price for Storage Capacity (metric units)

price of cistern ÷ storage capacity (liters) = price of storage capacity (price/liter)

See notes with Equation 11A. For non-USA currencies, substitute the appropriate currency.

Equation 12A.
Weight of Stored Water (English units)

stored water (gal) × 8.34 lb/gal = weight of stored water (lb)

EXAMPLE:

A 55-gallon drum under a rainspout has filled to the very top with water and you need to figure out how much it weighs to decide whether you can move it.

55 gal × 8.34 lb/gal = 458.7 lb

458.7 lb = weight of stored water

Water is extremely heavy. Do not underestimate the force you are dealing with when you store it. Platforms supporting storage tanks must be able to hold the water's weight!

Equation 12B.
Weight of Stored Water (metric units)

1 liter of water weighs 1 kilogram

So:
stored water (liters) × 1 kg/liter = weight of stored water (kg)

Appendix 4

Example Plant Lists and Water Requirement Calculations for Tucson, Arizona

This Appendix contains estimated water needs for vegetable gardens and three multi-use perennial plant lists specific for Tucson, Arizona (water needs will fluctuate depending on planting density, soil type, placement, and exposure). There is a far more diverse array of suitable plants and cultivars available for this area than the lists suggest. These lists are meant simply as both a partial introductory guide for Tucsonans, and as a template for people elsewhere to create plant lists specific to their location and climate.

Estimated annual or monthly water requirements can be easily calculated for plants by looking up their mature size, water needs (low, medium, high), and evergreen or deciduous nature on the plant lists, and then using the simple calculations that follow the lists. These estimates are very helpful in determining what plants, and how many, can be sustained within a Tucson, Arizona site's rainwater budget (calculated in chapter 2) and potential supplementary water from household greywater (estimated from box 2.6).

The vegetation section of the resources appendix of volume 2 of *Rainwater Harvesting for Drylands* lists some of the books from which I compiled the information. Local gardening groups, herbalists, primitive skills enthusiasts, native plant societies, locally owned plant nurseries, and my own direct observations then fleshed out the lists, and can help you form your lists too. Chapter 4 in this book, and the chapter on vegetation and the planting section of the chapter on infiltration basins in volume 2 offer still more tips.

The first table in box A4.1 shows, for various size vegetable gardens (square feet or square meters), approximate yearly water needs. Note that these gardens are mulched and in sunken basins, in conformance with the principles and strategies of water harvesting.

Box A4.1. Approximate Annual Water Requirements for Mulched Vegetable Gardens in Tucson, Arizona, Planted in Sunken Basins

Based on "Economic Value of Home Gardens in an Urban Desert Environment" by David A. Cleveland, Thomas V. Orum, and Nancy Ferguson, HortScience 20(4):694-696.1985

50 ft^2	100 ft^2	150 ft^2	200 ft^2	250 ft^2	300 ft^2
3,180 gallons	6,360 gal.	9,540 gal.	12,720 gal.	15,900 gal.	19,080 gal.
4.5 m^2	9 m^2	13.5 m^2	18 m^2	22.5 m^2	27 m^2
12,080 liters	24,160 liters	36,250 liters	48,080 liters	60,420 liters	72,500 liters

In the plant list tables that follow (boxes A4.2–A4.4), APPROXIMATE WATER NEEDS are listed as:

LW = low water use of 10 to 20 inches of water per year

MW = medium water use of 20 to 35 inches of water per year

HW = high water use of 35 to 60 inches of water per year.

The numbers 1, 2, 3, or 4 in parenthesis signify the approximate irrigation needs of the plants once they become established (this often takes 2 to 3 years).

(1) = no supplemental irrigation,

(2) = irrigation once a month in the growing season,

(3) = irrigation twice a month in the growing season,

(4) = irrigation once a week in the growing season.

Ratings based on Arizona Department of Water Resources Low Water Use/Drought Tolerant Plant Lists and direct observation.

Abbreviations signify: D=deciduous, E=evergreen, EO=essential oil, EPS=earth plaster/pigment stabilizer, F=food, FB=firebreak species, FR=fragrant, FW=fiber/basketry/weaving material, G=glue, H=hardy, HC= hair conditioner, LF=living fence, M=medicinal, NF=nitrogen-fixing, P=pigment or dye, S=shelter/shade, SC=screen, SD=semi-deciduous, SH=semi-hardy, SP=soap, T=tanning hides, W=wood/timber, WB=windbreak.

"Pollinators" can include: butterflies, native solitary bees, beneficial predatory wasps.

Box A4.2. Native Multi-Use Trees for the Tucson, Arizona Area

Species	Water	Size	Cold Tolerance	Elevation Range	Growth Rate	Type of Tree	Human Uses	Wildlife	Domestic Animals That Use Plant
Desert Ironwood (*Olneya tesota*)	LW (1)	25 × 25'	SH 15°F	2,500' and below	moderate	E	F, M, NF, S, T,	Birds, pollinators, large and small mammals	Chickens, goats
Velvet Mesquite (*Propsopis velutina*)	LW (1)	30 × 30'	H 5°F	1,000–5,000'	fast	SD	F, FW, M, NF, P, S, W	Birds, pollinators, large and small mammals	Chickens, goats, cattle, honey bees, dogs
Screwbean Mesquite (*Prosopis pubescens*)	LW (2–3)	20 × 20'	H 0°F	4,000' and below	moderate	D	F, FW, M, S, W, WB	Birds, pollinators, large and small mammals	Chickens, goats, cattle, honey bees, dogs
Cat claw Acacia (*Acacia greggii*)	LW (1)	20 × 20'	H 0°F	Below 5,000'	moderate to fast	D	M, P, S, T, W	Birds, pollinators, large and small mammals	Cattle, honey bees
Whitethorn Acacia (*Acacia constricta*)	LW (1)	10–15 × 10–15'	H 5°F	2,500–5,000'	moderate to fast	SD	F, G, M, S	Birds, pollinators, large and small mammals	Cattle
Desert Willow (*Chilopsis linearis*)	LW (2–3)	25 × 25'	H –10°F	1,500–5,000'	fast	D	FR, FW, M, S, W, WB	Birds and pollinators	Cattle, honey bees
Canyon Hackberry (*Celtis reticulata*)	MW (2–3)	Up to 35 × 35'	H–20°F	1,500–6,000'	moderate	D	F, S, W, WB	Birds, pollinators, large and small mammals	Chickens
Foothills Palo Verde (*Cercidium microphyllum*)	LW (1)	25 × 25'	H 15°F	500–4,000'	slow to moderate	D	F, S, W	Birds, pollinators, large and small mammals, desert tortoise	Cattle, honey bees
Blue Palo Verde (*Cercidium floridum*)	LW (2)	30 × 30'	H 15°F	500–4,000'	fast	D	F, S, W	Birds, pollinators, large and small mammals, desert tortoise	Sheep, honey bees

Box A4.3. Native Multi-Use Shrubs, Cacti, And Groundcover for the Tucson, Arizona Area

Species	Water	Size	Cold Tolerance	Elevation Range	Growth Rate	Type Of Plant	Human Uses	Wildlife	Domestic Animals That Use Plant
Oreganillo (*Aloysia Wrightii*)	LW (2)	5x5'	H 15°F	1,500–6,500'	moderate	D shrub	F, FR	pollinators	Honey bees, livestock
Quail-brush (*Atriplex lentiformis*)	LW (1)	Up to 8x12'	H 15°F	Below 4,000'	fast	E shrub	F, FB, M, NF, SC, SP	Birds, large mammals	Honey bees, livestock
Chiltepine (*Capsicum annum*)	LW (2)	Up to 3'	Frost sensitive	Below 4,000'	Slow to moderate	E shrub, D w/ frost	F, M	Birds	Chickens
Desert hackberry (*Celtis pallida*)	LW (2)	Up to 10'	H 20°F	1,500–3,500'	Slow to moderate	SD shrub	F, M, SC, W	Birds, pollinators, mammals	Chickens, Honey bees, cattle
Brittlebush (*Encelia farinosa*)	LW (1)	3'	SH 28°F	Below 3,000'	fast	E shrub	M, G	Pollinators, birds, large mammals	
Mormon Tea (*Ephedra trifurca*)	LW (2)	3–12'	H	Up to 4,500'	Slow	E shrub	E, M, P, T	Pollinators, birds, large mammals	Honey bees
Ocotillo (*Fouquiera splendens*)	LW (1)	Up to 15' tall	H 10°F	Below 5,000'	slow	D "shrub"	E, M, LF	Pollinators, birds	
Chuparosa (*Justicia californica*)	LW (2–3)	4'	SH 28°F	1,000–2,500'	Moderate to fast	D shrub	F	Birds, pollinators	
Creosote (*Larrea tridentata*)	LW (1)	Up to 11'	H 5°F	Below 4,500'	Slow to moderate	E shrub	G, M, W	Birds, pollinators, mammals	
Wolfberry (*Lycium fremontii*)	LW (1)	3–5'	H	2,500' and below	Moderate to fast	D shrub	F, M, SC	Birds, pollinators	Chickens, honey bees, livestock
Penstemon (*Penstemon parryi*)	LW (1)	Up to 3' tall	H 15°F	1,500–4,500'	Moderate	E ground cover	M	Birds, pollinators	
Jojoba (*Simmondsia chinensis*)	LW (1)	Up to 7'	H 20°F	1,000–5,000'	Slow to moderate	E shrub	FB, M, SC, SP, WB	Large and small mammals	Cattle
Saguaro (*Carnegiea gigantea*)	LW (1)	Up to 40' tall	SH 21°F	600–3,600'	slow	E cactus	F, G, M, W, T	Birds, bats, pollinators	Chickens
Barrel Cactus (*Ferocactus wislizenii*)	LW (1)	4–8' tall	H 15°F	1,000–5,600'	slow	E cactus	F, HC, M, P	Birds, pollinators, mammals	Pigs
Staghorn Cholla (*Opuntia versicolor*)	LW (1)	3–10' tall	H	2,000–3,000'	Moderate to fast	E cactus	F, M, SC	Birds, pollinators, mule deer	
Prickly Pear (*Opuntia Engelmanii*)	LW (1)	Up to 5' tall	H 10°F	1,000–6,500'	moderate	E cactus	EPS, F, LF, M, P	Birds, pollinators, mammals, tortoise	Sheep, cattle (when thorns burned off)

Box A4.4. Exotic Multi-Use Fruit Trees, Vines, and Cacti for the Tucson, Arizona Area

Species	Cultivars	Water	Size	Cold Tolerance or Needs	Growth Rate	Type of Plant	Human Uses	Wildlife	Domestic Animals That Use Plant
Apple (*Malus pumila*)	Anna, Ein Shemer	MW (3)	15–20' X 15–20'	150–250 chill hours	moderate	D tree	F, S	Birds, pollinators, deer	Chickens
Apricot (*Prunus armeniaca*)	Royal or Blenheim, Katy	MW (2–3)	25 X 25'	300–400 chill hours	moderate	D tree	F, FB, S, WB	pollinators	Chickens
Carob (*Ceratonia siliqua*)	Casuda, Santa Fe, Sfax	MW (3)	25 X 25'	SH 23°F	moderate	E tree	F, FB, S, WB,		Honey bees, sheep, goats, pigs, cows, horses
Chinese Jujube (*Ziziphus jujuba*)	Lang, Li	LW (2)	20–30 X 10–20'	H 0°F	moderate	D tree	F, M		Chickens
Citrus – grapefruit	Duncan, Ruby Red, Marsh	MW (3)	14–20'	SH 27°F	moderate	E tree	EO, F, FB, M, S	pollinators	Honey bees
Citrus – lemon	Improved Meyer, Lisbon	MW (3)	Up to 20 X 20'	SH 31°F	moderate	E tree	EO, F, FB, M, S	pollinators	Honey bees
Citrus – Sweet orange	Valencia, Trovita, Marrs, Sanguinelli Blood	MW (3)	12–20 X 12–20'	SH 27°F	moderate	E tree	EO, F, FB, FR, M, S	Pollinators, hummingbirds	Honey bees
Date palm (*Phoenix dactylifera*)	Medjool, khadrawy, halawy, zahidi, maktoom. Only females produce fruit	MW (3–4)	Up to 40' tall	SH 22°F	moderate	E tree	F, FW, M, S, W, WB	birds	Chickens, dogs, camels, horses
Grape (*Vitis spp.*)	Flame, Ruby, Lomanto, Black Manukka, Thompson	MW (4)	5–90' long	H 0–10°F	moderate	D vine	F, FW, S (on trellis)	Birds, pollinators, small mammals	Chickens, Honey bees
Fig (*Ficus carica*)	Black Mission, Conadria	MW (3)	15–30 X 15–30'	H 15°F >100 chill hours	fast	D tree	F, FB, M, S	Birds, bats, pollinators	Chickens
Loquat (*Eriobotrya japonica*)	Big Jim, Tanaka, Champagne, Gold Nugget	HW (4)	20 X 20'	Tree H 10°F , fruit & flowers SH 28°F	moderate	E tree	F, S, WB		Chickens, honey bees
Nopal (*Opuntia ficus-indica*)	Burbank, Quillota, Papaya, Honey Dew, Florida White	LW (1–2)	Up to 10' tall	H 20°F	moderate–fast	E cactus	EPS, F, FB, LF, M, SC	Pollinators, desert tortoise, javalina	Chickens, pigs, sheep, cattle
Olive (*Olea europaea*) *	Ascolano, Barouni, Haas, Manzanillo, Mission	MW (2–3)	Up to 30 X 30'	Trees H 15°F, Green fruit SH 28°F	moderate	E tree	M, S, W, F, FB, WB	Birds	Chickens
Peach (*Prunus persica*)	Desert Gold, Mid Pride, Rio Grande	MW (3–4)	15–25'	H –15°F , 250–350 chill hours	moderate to fast	D tree	F, FB, M, S	Birds, pollinators	Chickens, honey bees
Pomegranate (*Punica granatum*)	Wonderful, Fleishman, Papago, Sweet	LW (2–3)	12–15'	H 15°F, 100–200 chill hours	moderate	D shrub to tree	F, FB, M, P, SC, T	Birds	Chickens, honey bees

* Order fruiting olives from Peaceful Valley Farm Supply (www.groworganic.com).

How to Estimate the Water Requirements **in a Given Month** for a Listed Plant in Tucson, Arizona

Based on the "How To Develop A Drip Irrigation Schedule" handout from the LOW 4 Program of the Pima County Cooperative Extension/University of Arizona Water Resource Research Center 350 N. Campbell Ave., Tucson, AZ 85719. Ph. 520-622-7701.

A similar "plant water requirement estimator" can be created for other areas according to local evapotranspiration rates.

For an additional resource, see the Arizona Department of Water Resources for their Drought Tolerant/Low Water Use Plant Lists http://www.water.az.gov/adwr/Content/Conservation/LowWaterPlantLists/default.htm

They have plant lists specific to Tucson, Phoenix, and the Pinal, Prescott, and Santa Cruz Active Management Areas (AMAs)

1. Identify the plant as evergreen or deciduous, and as high, medium, or low water requirement.

For example, a Velvet Mesquite is deciduous with a low water requirement.

2. Determine the canopy diameter of the plant (the diameter of the leafy part of the plant). This can be the plant's current canopy or its potential canopy at maturity. *Let's say our example mesquite has a 20-foot canopy.*

3. Determine the plant's water requirement in inches for a given month. See the tables in boxes A4.5A and A4.5B, which show how many INCHES of water the plant needs to receive beneath its canopy to maintain its health. *According to the table in box A4.5B, the June water requirement of our deciduous, low water requirement mesquite is 3 inches.*

4. Convert the plant's water requirement from inches to gallons. Find the plant's canopy diameter in Box A4.5C. Then find the corresponding # of gallons per inch of water beneath the canopy, and multiply it by the number of inches required in June to get the total GALLONS of water required in that month. *For example, the number of gallons in an inch of water under a 20-foot diameter Velvet Mesquite is 196 gallons. The tree needs 3 inches of water in June, so multiplying 196 × 3 = a June water requirement of 588 gallons.*

Box A4.5A. Monthly Water Requirement in Inches–Evergreen Plants

Water Requirement	J	F	M	A	M	J	J	A	S	O	N	D	Annual Total
Low	0	0	2"	2"	3"	3"	3"	2"	2"	2"	1"	0	20"
Medium	0	0	3"	4"	5"	5"	5"	4"	4"	3"	2"	0	35"
High	0	3"	5"	6"	8"	9"	7"	6"	6"	5"	3"	0	58"

Box A4.5B. Monthly Water Requirement in Inches–Deciduous Plants

Water Requirement	J	F	M	A	M	J	J	A	S	O	N	D	Annual Total
Low	0	0	0	2"	3"	3"	3"	2"	2"	0	0	0	15"
Medium	0	0	0	4"	5"	5"	5"	4"	4"	0	0	0	27"
High	0	0	0	6"	8"	9"	7"	6"	6"	5"	0	0	47"

Box A4.5C. Conversion Table: Canopy Diameter vs. Gallons/Inch under Canopy											
Canopy Diameter in Feet	2	4	6	8	10	12	14	16	18	20	25	30
# of Gallons per Inch of Water beneath Canopy	2	8	18	31	49	71	96	125	159	196	306	441

How to Estimate the Annual Water Requirements for a Listed Plant in Tucson, Arizona

Use the tables in box A4.5A or A4.5B to find the plant's estimated ANNUAL water requirement in INCHES. Multiply that number by the number of gallons per inch of water beneath the canopy (table in box A4.5C), and the plant's canopy diameter. *For example, the 20-foot diameter Velvet Mesquite needs 15 inches of water annually, and from Table A4.5C we see that there are 196 gallons per inch of water beneath a 20-foot canopy. So multiplying 15 × 196 = an annual water requirement of 2,940 gallons.*

Note 1: Annual water requirement estimates are likely all you will need to consider when designing a landscape of local native plants based on natural wild plant densities and sizes. Such vegetation is naturally adapted to the local rainfall patterns and, once established, can survive the dry periods between rains.

Monthly water requirement estimates are better suited for designing landscapes of exotics or native plants that are planted at a higher than normal density or are irrigated for larger than normal plant sizes. These estimates give you a better idea of what seasons or months require more water so that you can better plan for needed water storage and the timing of supplemental irrigation with cistern water or greywater.

The water requirements for all plants will increase as they grow, since the amount of water they transpire through their leaves increases with the increase in cumulative leaf surface area. Therefore, it is important to plan for the water needs of your plants at their mature size. However, by minimizing the amount of water available to native plants you can reduce their mature size—reducing the need for more water. For example, a Velvet Mesquite receiving approximately 6,600 gallons of water per year can grow to be 30 feet tall and wide, but if only 2,940 gallons of water per year is available to the tree, it will likely not grow to be taller and wider than 20 feet.

Note 2: For another method of estimating landscape water needs and tables of information allowing you to do so for many locations in Arizona see the free publication, *Harvesting Rainwater for Landscape Use*, 2nd Edition by Patricia H. Waterfall and Christina Bickelmann, 2004. The document may be ordered from the Arizona Department of Water Resources, Tucson Active Management Area, 400 W. Congress, Suite 518, Tucson, AZ 85701, Phone 520-770-3800, Website www.water.az.gov. This document is also available online: www.cals.arizona.edu/pubs/water/az1344.pdf

Appendix 5
Worksheets: Your Thinking Sheets

This appendix is meant both as a *summary of the design process* described in this book and as a checklist of the recommended steps toward creating an integrated conceptual water harvesting design. It's intended to help you determine and lay out what strategies should be used on your site and where to maximize potential and efficiency of your site as a whole. You may find it helpful to refer back to the chapters from which the various steps are taken for elaboration.

Consider photocopying these worksheets—and especially any site maps you may have drawn, so you can record revisions and incremental layers of your insights—and mark up the copies rather than the book. Feel free to add sticky notes and additional pages of notes and revisions. Please be aware that this material is copyrighted. You may **photocopy** these pages of this appendix for your own personal use and that of members of your household **only**. You may not copy these sheets to a computer. Please contact Brad Lancaster for permission for all other uses, and other media (computer, cd-rom, etc.).

This appendix follows the thought-flow in volume 1 of *Rainwater Harvesting for Drylands*. The result will be a series of sheets, beginning with any thoughts you may have on applying the principles of rainwater harvesting to your site, to basic observations about your land, and leading to steps you can take to have a more water-abundant home and yard—one more comfortable and beautiful in both winter and summer.

Please don't go overboard trying to make beautiful or accurate site maps, finding all the information at first, or measuring everything or performing all the calculations (though to maximize the results of your rainwater-harvesting set-up the measurements and calculations will eventually have to be done accurately). As you find you need information, research it and fill it in later. This is an ongoing process, and should be a one of adventurous discovery.

CHAPTER 1/STEP 1: THE RAINWATER-HARVESTING PRINCIPLES

Here's a summary of the "Principles of Successful Rainwater Harvesting" found on pages 29–38. Add any of your thoughts below them. Continually refer to these principles as you assess you site, envision your water harvesting approach, and implement the design.

The Principles of Successful Rainwater Harvesting

1. Begin with long and thoughtful observation.

Use all your senses to see where the water flows and how. What is working, what is not? Build on what works.

2. Start at the top (highpoint) of your watershed and work your way down.

Water travels downhill, so collect water at your high points for easy gravity-fed distribution. Start at the top where there is less volume and velocity of water.

3. Start small and simple.

Work at the human scale so you can build and repair everything. Many small strategies are far more effective than one big one when you are trying to infiltrate water into the soil.

4. Spread and infiltrate the flow of water.

Rather than having water erosively runoff the land's surface, encourage it to stick around, "walk" around, and infiltrate *into* the soil. Slow it, spread it, sink it.

5. Always plan an overflow route, and manage that overflow as a resource.

Always have an overflow route for the water in times of extra heavy rains, and where possible, use that overflow as a resource.

6. Maximize living and organic groundcover.

Create a living sponge so the harvested water is used to create more resources, such as cooling shade and food. As plants grow the soil's ability to infiltrate and hold water steadily improves.

7. Maximize beneficial relationships and efficiency by "stacking functions."

Get your water harvesting strategies to do more than hold water. Berms can double as high and dry raised paths. Plantings can be placed to cool buildings. Vegetation can be selected to provide food.

8. Continually reassess your system: the "feedback loop."

Observe how your work affects the site—beginning again with the first principle. Make any needed changes, using the principles to guide you.

INTERLUDE: YOU AND YOUR WATER

How do you think you currently use your water resources (rough breakdown)?

    indoors_____

    outdoors_____

Do you wish you had:

    a cleaner or better source of water?_____

    less expensive water?_____

    a more direct connection with your water source and water use?_____

Have you ever thought of conserving water?_____

    If so, how?_____

If you had a better source of water or less expensive water, how would you use it?

    landscape and garden use_____

    washing and bathing_____

    potable use (drinking and cooking)_____

    other_____

After reading/perusing this book, what might you want to do?

    first?_____

    second?_____

    third?_____

CHAPTER 2/STEP 2: BUILDING ON LONG AND THOUGHTFUL OBSERVATION—ASSESS YOUR SITE'S WATERSHED AND WATER RESOURCES.

A. Walk your site's watershed.

Identify its ridgelines/boundaries, and observe how water flows within them. Make any notes below. Refer to pages 42–44 for more information.

If runoff flows across your land, pay particular attention to what direction it comes from, its volume, and the force of the water's flow. Look at what surfaces water flows over to estimate the water's quality. Note any observations below. You may also want to look for erosion patterns when it's dry (see appendix 1 on page 111). Write down your observations.

B. Create a site plan and map your observations.

First, photograph your site so you'll have "before" photos with which to document your progress with future "after" photos. You may want to include copies of these photos with these worksheets.

Now, use the grid paper at the end of this appendix (you can purchase your own if desired) and using figure 2.3 on page 47 as a model create a "to-scale" site plan of your property's boundaries. Leave wide margins to mark the locations where resources—such as runoff from your neighbor's yard—flow on, off, or along-side your site. Draw buildings, driveways, patios, existing vegetation, natural waterways, underground and above-ground utility lines, etc. to scale. Next, draw any catchment surfaces that drain water off your site (for example, a driveway sloping toward the street), and any catchment surfaces draining water onto your site from off-site; indicate the direction and flow of any runoff and runon water. Refer to pages 44–50 as you do all this. Write down additional notes below.

Additional observations you may want to record at this time:

What vegetation lives solely off on-site water (rainwater), and which depends on pumped water or imported irrigation water?

What unirrigated native vegetation do you see growing within a 25-mile (40-km) radius of your site that could do well on your site?

C. Calculate your site's rainwater resources.

C1. Your Site's Rain "Income"

Determine the "income" side of your site's water budget so you can compare them with the "expense" side. For this section refer to pages 44–51 and additional calculations found in appendix 3.

- What is the area's average annual precipitation in inches or mm?_____
- What is the area of your site (land) in square feet, acres, or hectares?
- What is the area of the roofs of your house, garage, sheds, and other buildings on your property (see the calculations appendix 3, equations 1–3 on pages 126–127)?

_____ _____ _____

Now, use the calculations on pages 45 and 48, or in appendix 3, to determine your site's rainfall resources. You will want to answer the question: In an average year, how much rain falls on your site? For this you will use your site's area and annual precipitation to get some kind of ballpark figures about your annual rainwater resources in gallons or liters. If you have a difficult time with the math, then just use the "rule of thumb" figures in appendix 3 on page 129. Do your calculations below.

X. Your site's annual rainfall resources in gallons or liters. How much water falls onto your site?

X = _____

C2. Your Site's Rain "Loss"

Now, refer to the calculations on pages 45 and 48 to determine how much runoff drains from your impervious catchment surfaces for potential storage/use in adjoining tanks or earthworks. You don't need to be exact; you just want a good estimation.

Roof runoff_____
Driveway runoff_____
Patio runoff_____
_____ runoff_____

Then note how much of that runoff (and additional potential runoff from other built or disturbed surfaces such as mounded sections of the landscape or bare dirt areas) currently drains off your property. Add the estimated total and write it below:

Y. Loss/runoff from your site in gallons or liters. How much runoff runs off your site?

Y = _____

C3. Your Site's Water Gain

Now, you want to estimate how much water (annually) you're gaining from runoff from other properties onto your site. Use the same calculations as above.

Z. Gain/runon in gallons or liters. How much off-site runoff runs onto your site? (same calculations as above)

Z = _____

C4. Totaling It Up

X (on-site rainfall) - Y (runoff draining off site) + Z (runon to your site) = T (TOTAL: HOW MUCH YOU CURRENTLY DO HARVEST)

X_____ - Y_____ + Z_____ = T_____

T = _____

This equals the total on-site rainwater resources you currently do harvest.

C5. Your POSSIBLE Rainwater Harvest

Now, exclude Y, your debits, i.e., the rainwater that is runoff from your site:

X (on-site rainfall) + Z (runon to your site)= TT (TOTAL POSSIBLE RAINWATER HARVEST)

X_____ + Z_____ = TT_____

TT = _____

This may illustrate how much more water you *could* harvest with the use of various water-harvesting earthworks or storage techniques.

Write down any additional observations and perform any calculations. Compare how much water you *do* harvest (T) to that which you *could* harvest (TT).

D. Estimate your site's water needs.

This step determines the "expense" side of your water budget by estimating your household and landscape water needs. See pages 51–53.

- What is your annual water consumption based on your water bill?_____
- In which months are your water consumption/needs highest?_____

The next steps are to try to determine how much of your water is used indoors versus outdoors.

- Estimate your average annual *indoor* water consumption using the user-friendly website www.h2ouse.org (see pages 51–53)._____

- Estimate your average annual *outdoor* water consumption (based on plant water-need requirements; see appendix 4 on page 136): List some of your larger plants and their water requirements below.

- Or/And: Subtract your estimated indoor water consumption from your water bill for a ballpark estimate of your current outdoor needs._____

E. Compare your needs and resources.

Compare your site's water needs to the volume of its rainwater *resources* on-site or flowing onto it. Review the information you've gathered and calculated in section C above; also refer to chapter 3 and the calculations found in appendix 3 as needed. This section involves recording any thoughts you may have about how to balance your site's water budget and also perhaps doing some additional calculations.

- How much water *could* you harvest on-site? (TT from p.145)_____

- How much of your domestic water needs could you meet by harvesting *rooftop runoff* in one or more tanks? _____

- How much vegetation could you support by simply harvesting rain falling and infiltrating directly in your soil?_____

- How much vegetation could you support if on-site runoff was also directed to the planted areas (this runoff could be diverted directly and passively to planted areas or harvested in a tank and doled out to the planted areas as needed)?

- List what other steps you could take to balance your water budget using harvested rainwater as your primary water source. See the conservation strategy suggestions in box I.7 on page 13.
Indoors_____

Outdoors_____

F. Greywater sources

Estimate the average volume of accessible household greywater you could reuse within your landscape, using the information in box 2.6 on page 53 or from www.greywater.com. Accessible means you can access current drain pipes or install new ones to direct the greywater to mulched and vegetated basins within the landscape. You will need to maintain a minimum 1/4 inch drop per linear foot of pipe (2 cm drop per linear

meter) for gravity to freely and conveniently distribute your greywater from a point downstream of the p-trap for the greywater source (washing machine, sink, etc.) to the greywater pipe outlet in the landscape.
washing machine_____
shower_____
bathtub_____
bathroom sink_____
other_____
Total _____

CHAPTER 3/STEP 3: EARTHWORKS, TANKS, OR BOTH

Refer to the comparisons in box 3.1 (page 57) and see the overview of strategies later in chapter 3 to decide how you might best harvest the water for your planned uses.

Now that you've (more or less) estimated your on-site water resources and needs (from the previous worksheets), the next step is to answer again the following questions:

How do you *currently* use your water resources (rough breakdown):
indoors_____
outdoors_____

How do you *plan* to use your water resources?
landscape and garden use_____
washing and bathing_____
potable use_____
other_____

After reviewing chapter 3, and the "principles of successful rainwater harvesting" from chapter 1, what do you think you'd want to do:
first?_____
second?_____
third?_____

Compare the above to your answers in the earlier interlude. What's different, now that you have more information to work from?

CHAPTER 4/STEP 4: INTEGRATED DESIGN

This chapter is intended to heighten your awareness of additional on-site resources and challenges, and to show you how to maximize their potential by integrating their harvest with that of water. The numbers below follow the Integrated Design Patterns and their Action Steps found on pages 81–101.

Make a new photocopy of your site map and mark the directions of north, south, east, and west.

1. Your site's orientation

(Consult pages 81–101; see especially the figures which can be used as models for marking your own site map.)
- What is your site's latitude (ask your local friendly librarian if you don't know):_____
- How is your site and/or your home oriented? (Use a compass or the sun's orientation; see page 81). Put this information on your site map as well as writing it below.

On your site map:
- Identify the "winter-sun side," and the "winter-shade side" of your home.
- Map the location of the rising and setting sun on the summer and winter solstice.
- Also mark any of the following, and any additional incoming resources or challenges (see figure 4.4, page 82): where you would like more shade or exposure to sun; the direction or location where prevailing winds, noise, or light come from; and the foot traffic patterns of people, pets, or wildlife.
- Where are the warm and cold spots in winter? The hot and cool spots in summer? What areas inside your house get direct sun in the morning and afternoon, in winter and summer? Do you get sun and shade on your garden and outdoor areas when and where you want it? Mark your site map appropriately and write your observations below.

- Shut off mechanical heating and cooling systems at least once in each season of the year to observe how direct solar exposure—or the lack of it—affects the comfort of your home and yard. When you do this, what are your observations?

2. Window overhangs (pages 85–88)

- Do you have window overhangs?_____
- If so, what is their projection length on the *winter-sun side* of your house?_____
- Use the overhang projection information on pages 85–87 and box 4.3 to determine appropriate overhang sizes for your winter-sun side windows._____
- Compare existing overhangs to what the calculation recommends._____
- What have you noticed about how overhangs or the lack of them affect your comfort throughout the year?_____
- What can you do (put up awnings or trellises, plant trees, extend overhangs, open up a covered section of winter-sun-facing porch, etc.) to enhance the positive ways sun and shade can affect your building?_____

3. Your house in the bigger picture: solar arcs (pages 88–90)

Make a new site map if needed.

- Do you have any elements of a solar arc in place around your home or garden, such as an existing shade tree, covered porch, or building? If so, mark them on your site map and write comments about them below.

- Now, indicate on your site map where missing pieces of a solar arc should be located to complete it and benefit your home or garden.

- Can you use any water-harvesting strategies (earthworks, trees, cisterns) to create or grow a solar arc or a windbreak?

4. Sun traps (pages 90–92)

On your site map, mark where a sun trap might make sense, and indicate any existing elements already in place. Write your observations below.

- Consider the desirability of a fence, new cistern, or plantings within earthworks such as trees, large shrubs, or vines growing on trellises, fences, etc. to create a sun trap. Write any thoughts below.

5. Maintaining winter sun exposure (see pages 93–95)

- Where have you noticed winter shadows blocking your sun? Write down observations below, and on your site map indicate features (trees, etc.) that produce long winter shadows.

- What is your latitude?_____
- What would be the length of a shadow cast by a 20-foot tree by the noonday sun at your latitude (see the shadow-ratio correlation in box 4.7 on page 93)?_____
- Now, think about where you might want to add appropriately placed new vegetation, structures, or windbreaks to avoid blocking desired winter sunlight for winter sun-facing windows, winter gardens, and solar strategies, while providing other benifits. Note any observations below, and pencil in on your site map if necessary.

- Are there any features you'd want to remove or relocate that block winter sunlight?_____
- Add any additional comments below.

6. Raised paths, sunken basins—relative height (pages 95–96)

- Write your observations of the following: the relative height of paths, patios, sidewalks, driveways and streets compared to adjacent planting areas in your home and community.

- Do you see any "raised path, sunken basin" patterns or a sunken path, raised planting area pattern?

- Is stormwater being directed to vegetation, asphalt, or storm drains?

- Now, identify and map areas where you *could* develop the raised path, sunken basin pattern at home.

7. Reducing paving and making it permeable (pages 96–99)

- Below, write examples of pervious and impervious paving around your home and community.

- Write any thoughts about how you can: reduce the paving on your site; either direct the remaining pavement's runoff into adjoining earthworks or make remaining pavement more permeable; turn your driveway into a "park-way" or use porous brick, cobbles, or angular open-graded gravel instead of an impervious material such as concrete; etc. (Refer also to pages 64–66 for more ideas and read volume 2, the chapter on reducing hardscape and creating permeable paving.)

8. Tying it all together—creating an integrated conceptual design (pages 100–101)

Now, you can really use the information and insights gained so far. Focus now on placing and integrating the various elements on your site, while referring to volumes 2 and 3 for specifics.

Again make multiple copies of your site plan.

Use these copies as base maps for sketching out draft designs of your site's integrated water-harvesting system. Continually refer to your mapped observations to see how your design ideas integrate with and can best build on what exists.

Play with different conceptual water-harvesting plan layouts. I recommend two options:

- Make cut-outs (in the same scale as your site plan, and perhaps using sticky paper or sticky notes) of the trees, cisterns, patios, gardens, and other elements you want to add to your site. Move these cut-outs around your site plan imagining how they will interact with the flows of your on-site resources (rainwater, greywater, sun, wind, etc.).
- Put tracing paper over your site plan and sketch where you could place various elements (trees, cisterns, patios, gardens, and other elements) you want to introduce to your site, and then see how they interact with on-site resource flows.

As you keep playing with various arrangements, ask yourself, "Where do I need water, where do I have it, how much do I have, and how/where can I best utilize it?" Remember, your goal is to increase site efficiency and maximize site potential. Write down any additional observations below.

And again . . . before you dig (or order that cistern)—

Refine your design by planning the water-harvesting details or calculating a planned cistern(s) size. (The chapters in *Rainwater Harvesting for Drylands*, volumes 2 and 3, have all the needed information for such refinement.) Always remember to continually refer back to the rainwater-harvesting principles in chapter 1 to make sure they are all realized in your evolving plan.

Now, walk your land again, imagining how various strategies could work within the unique context of your site. What are your observations?

Play more with ideas and layouts on paper—it's much easier to make changes with a pencil and eraser than with a shovel. What have you changed?_____

When you feel you've got your plan set, scratch out, stake, or spray paint locations of paths, trees, water-harvesting strategies, and other elements in the dirt at your site. Walk around you site feeling what it's like to inhabit this system. Make any needed changes and if all feels good—go for it.

But, you don't have to implement everything all at once. Rather, use the rainwater-harvesting principles to prioritize your work—*start small, start at the top . . .* And—*if future observations or realizations justify a change in your plan after you've begun implementation, make a change.* Just add your new observations to those you've previously made on your site map, assess the best integrated approach, and move forward. This is all a process based on *long and thoughtful* observation, continuing for the duration of your relationship with the site.

APPENDIX 3: CALCULATIONS

Look through the equations. If you feel that any are useful to your situation, space is provided below for your calculations and notes.

⅛-inch squares

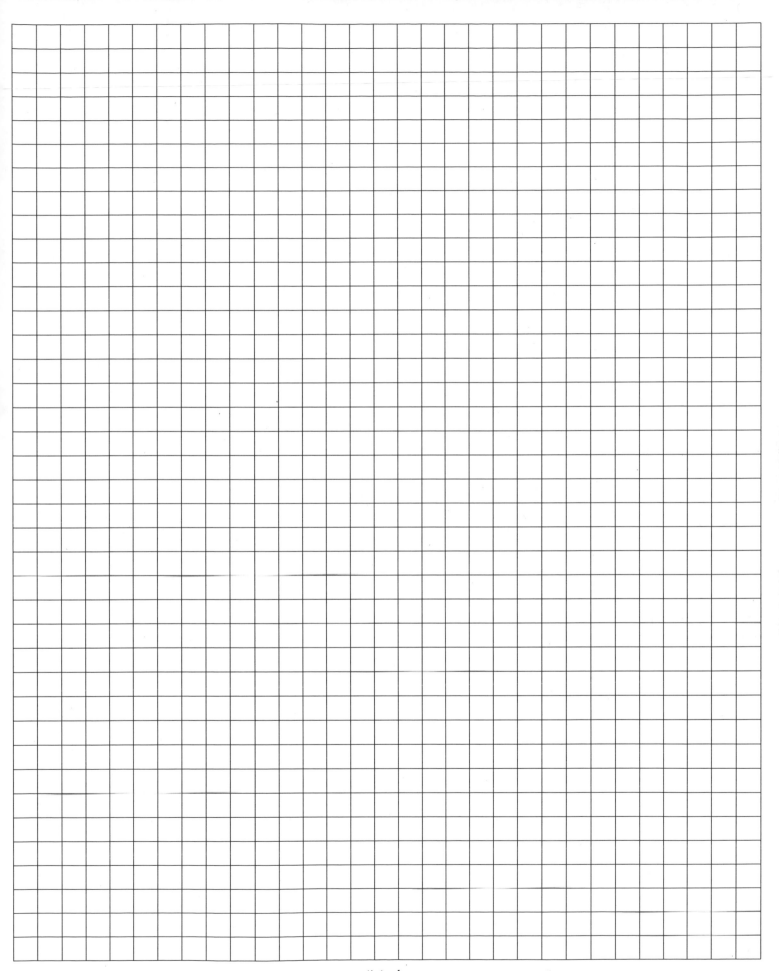

¹/₄-inch squares

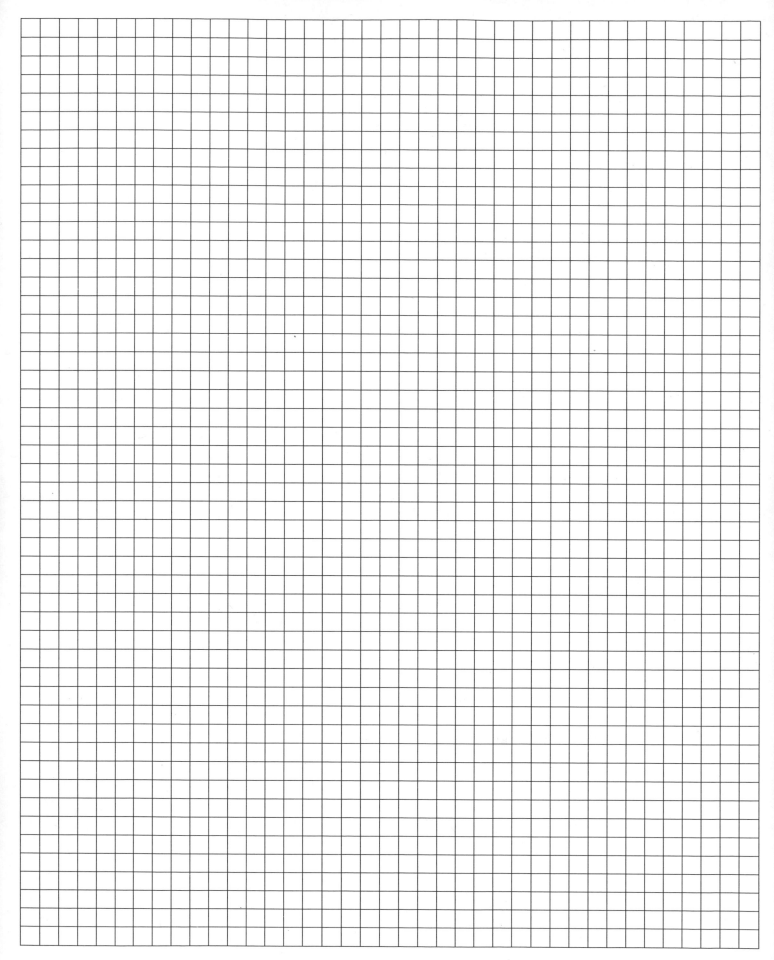

5mm squares

Appendix 6

Resources

This appendix provides a comprehensive list of helpful resources; it includes much more than just the texts cited in this volume. This list begins with general rainwater-harvesting sources, then follows the topical order in volume 1's introduction and chapters 1 through 4. Sections O through X provide helpful permaculture, community, government, and funding resources.

A. GENERAL RAINWATER HARVESTING RESOURCES

PUBLICATIONS

Rainwater Harvesting for Drylands, vol. 1: Guiding Principles to Welcome Rain Into Your Life & Landscape, by Brad Lancaster. Rainsource Press, www.HarvestingRainwater.com, 2006.

City of Tucson Water Harvesting Guidance Manual, edited by Ann Audrey Phillips. Available from City of Tucson, Department of Transportation, Stormwater Section, PO Box 27210, Tucson, Arizona 85726. Online: www.ci.tucson.az.us/water/harvesting.htm. A great guidance manual providing basic information and design ideas for developers, engineers, designers, and contractors of commercial sites, public buildings, subdivisions, and public rights-of-way.

Stormwater as a Resource: How to Harvest and Protect a Dryland Treasure, by David Morgan and Sandy Trevathan. A collaboration between the City of Santa Fe and the College of Santa Fe, 2002. Available from www.nmenv.state.nm.us/swqb/Storm_Water_as_a_Res ource.pdf. This booklet is a brief, clear, and concise guide for harvesting rain and snow on your property.

Harvesting Rainwater for Landscape Use, 2nd ed., by Patricia H. Waterfall and Christina Bickelmann. Cooperative Extension, College of Agriculture and Life Sciences, University of Arizona, 2004. Order from Arizona Department of Water Resources, Tucson Active Management Area, 400 W. Congress, Suite 518, Tucson, AZ 85701; phone: 520-770-3800; website: www.water.az.gov. Available online: www.cals.arizona.edu/pubs/water/az1344.pdf. Good basic guide with calculations for estimating water needs of landscape vegetation.

Forgotten Rain: Rediscovering Rainwater Harvesting, by Heather Kinkade-Levario. Granite Canyon Publications, 2004. Website: www.forgottenrain.com. This Arizona-focused guide has many detailed illustrations of various cistern and filtration systems, plus case studies of Arizona water harvesting sites.

A Water Harvesting Manual for Urban Areas: Case Studies from Delhi, from the Centre for Science and

Environment, www.cseindia.org, 2000. A very accessible guide encouraging community and household-based water harvesting in India.

Making Water Everybody's Business: Practice and Policy of Water Harvesting, from the Centre for Science and Environment, www.cseindia.org, 2001. A great book documenting numerous water-harvesting projects in India and around the world.

Dying Wisdom: Rise, Fall, and Potential of India's Traditional Water Harvesting Systems, edited by Anil Agarwal and Sunita Narain. Centre for Science and Environment, www.cseindia.org, 1997. A fascinating account of diverse water-harvesting systems throughout India.

The Negev: The Challenge of a Desert, 2nd ed., by Michael Evenari, Leslie Shanan, and Naphtali Tadmor. Harvard University Press, 1982. A study of ancient and recreated water harvesting and runoff agriculture in the Negev desert.

The Collection of Rainfall and Runoff in Rural Areas, by Arnold Pacey and Adrian Cullis. Intermediate Technology Publications, 1986. A dry, but informative resource with a worthy emphasis on recognizing local needs and utilizing local resources.

ONLINE RESOURCE (IN ADDITION TO ABOVE)

www.rainwaterharvesting.org.
This site belongs to the Centre for Science and Environment (CSE), one of India's leading environmental NGOs. Although its primary focus is on rainwater harvesting in India, there is much information pertinent to rainwater harvesting around the globe.

VIDEOS

Harvest of Rain. Centre for Science and Environment, www.cseindia.org, 1995. Traditional water harvesting systems of India are featured in this 48-minute video.

B (INTRODUCTION). WATER ISSUES AND PROTECTING THE RIGHT TO CLEAN WATER FOR ALL CITIZENS OF THE EARTH (INCLUDING WILDLIFE)

PUBLICATIONS

Killing the Hidden Waters: Slow Destruction of Water Resources in the American Southwest, by Charles Bowden. University of Texas Press, 1977. A well-written book on how various cultures in the Southwest U.S. have decided to use our water and other resources, and what effect this has had on the people and the environment.

Keepers of the Spring: Reclaiming Our Water in an Age of Globalization, by Fred Pearce. Island Press, 2004. An excellent resource documenting both the mistakes of inappropriate western engineering schemes that make fresh water scarcer, and the successes of indigenous traditional water harvesting schemes that lead to fresh water abundance.

Cadillac Desert: The American West and Its Disappearing Water, by Marc Risner. Penguin Books, 1993. A very well-written book on water policy, politics, and use in the American West. A video series based on the book is also available.

Water Follies: Groundwater Pumping and the Fate of America's Fresh Waters, by Robert Glennon. Island Press, 2002. A great book on the consequences of our country's growing dependence on our dwindling groundwater resources.

Desert Waters: From Ancient Aquifers to Modern Demands, by Nancy R. Laney. Arizona-Sonora Desert Museum, 1998. A short and concise publication on our water situation in the Southwest with tips on how to reduce water use.

Blue Gold: The Global Water Crisis and the Commodification of the World's Water Supply, by Maude Barlow. International Forum on Globalization Special Report, 1999. Online summary and ordering information at ifg.org/bgsummary.html. This is a very clear and concise report on the state of our fresh water resources and how we can protect and enhance them.

Blue Gold: The Fight to Stop the Corporate Theft of the World's Water, by Maude Barlow and Tony Clarke. The New Press, 2002. An important book on the world's growing fresh water crisis, the corporate assault on the water "commons," and how ordinary citizens all over the world are taking back control, becoming the "keepers" of the fresh-water systems in their localities.

Last Oasis: Facing Water Scarcity, by Sandra Postel. Worldwatch Institute, 1997. A great book looking at the mismanagement of the world's water resources, and how we can promote more sustainable use of water through conservation and a water ethic.

Water Wars: Privatization, Pollution, and Profit, by Vandana Shiva. South End Press, 2002. An excellent book examining the international water trade, damming, mining, and aquafarming, Shiva exposes the destruction of the earth and the disenfranchisement of the world's poor as they are stripped of their right to a precious common good.

"Troubled Waters," by Sandra Postel. *The Sciences* (March/April 2000). A look at the world's fragile supply of fresh water.

YES! A Journal of Positive Futures, no. 28, Winter 2004. This issue is devoted to water issues including access to fresh water as a human right, protecting watersheds, indigenous water conservation, and more.

The World's Water, 2004-2005, The Biennial Report on Freshwater Resources, by Peter Gleick. Island Press, 2005. Gives a global overview of water use by country, dams by country, etc.

ONLINE RESOURCES

www.citizen.org/cmep/water.
The Water for All Campaign of the national non-profit public interest organization, Public Citizen.

C (INTRODUCTION). BOOKS ABOUT THE PEOPLE AND WATER AND HOW THEY SHAPE EACH OTHER

PUBLICATIONS

The Secret Knowledge of Water: Discovering the Essence of the American Desert, by Craig Childs. Sasquatch Books, 2000. A wonderful book about the author's endless search for water in the desert. Incredible adventures lead to his discovery that the desert is nothing but water.

The Desert Smells Like Rain, by Gary Paul Nabhan. North Point Press, 1982. A graceful and humane tour of the Tohono O' Odham and how they live in the beautiful Sonoran desert.

Cultures of Habitat: On Nature, Culture, and Story, by Gary Paul Nabhan. Counterpoint, 1997. A wonderful book celebrating how people and nature can coexist and enhance one another.

ONLINE WATER FACTS

www.h2o4u.org/facts.html. Fascinating water facts.

ct.water.usgs.gov/EDUCATION/trivia.htm. More water facts.

D (CHAPTER 1). ON MR. ZEPHANIAH PHIRI, ZWRP, ZVISHAVANE WATER RESOURCES PROJECT

PUBLICATIONS

The Water Harvester: Episodes from the Inspired Life of Zephaniah Phiri, by Mary Witoshynsky. Weaver Press, 2000. Address: Weaver Press, Box A1922, Harare, Zimbabwe.

ORGANIZATIONS

If you'd like to support the great work of this grass-roots project write to: Mr. Zephaniah Phiri Maseko, ZWRP, P.O. Box 118, Zvishavane, Zimbabwe.

E (CHAPTER 1). PERMACULTURE, GENERAL WORKS

PUBLICATIONS

Introduction to Permaculture, by Bill Mollison. Tagari Publications, 1988. A smaller, more readable version of *Permaculture: A Designer's Manual* without the drylands emphasis.

Permaculture: A Designer's Manual, by Bill Mollison. Tagari Publications, 1988. The permaculture Big Book with a good emphasis on drylands, and an even better emphasis on how to incorporate water harvesting into an efficient, holistic system.

Gaia's Garden: A Guide to Home-Scale Permaculture, by Toby Hemenway. Chelsea Green Publishing Company, 2001. While not dryland-specific, this book clearly describes how you can work the integrated design of permaculture into your backyard garden and landscape.

Permaculture: Principles and Pathways Beyond Sustainability, by David Holmgren. Holmgren Design Services, 2002. A more cerebral and very worthwhile book based on the co-originator of permaculture's extensive practical experience, offering a deeper understanding of permaculture concepts.

MAGAZINES

The Permaculture Activist. Subscription information at www.permacultureactivist.net.

VIDEOS

Global Gardener. Bullfrog Films, 1991. Order from: Bullfrog Films, P.O. Box 149, Oley, PA 19547;

phone 1-800-543-3764. Travel the world to see how permaculture approaches to sustainable agriculture have turned wastelands into food forests in the tropics, drylands, cool climates, and the urban environment.

F (CHAPTER 2). WATERSHED RESOURCES

ONLINE RESOURCES

cfpub1.epa.gov/surf/locate/map2.cfm. "Surf Your Watershed," a website of the Environmental Protection Agency (EPA). Here you can locate the regional watershed of which your town or site is a part, and get information about that watershed, though this website lacks the detail to show you the boundaries of smaller watersheds.

MAPS AND AERIAL PHOTOGRAPHS

mapping.usgs.gov. United States Geological Survey (USGS) topographic contour maps, sometimes called "topo maps," can be very helpful in determining watershed boundaries since they illustrate the changing elevation of a landscape.

Urban environments: You can often obtain detailed topo maps, or aerial photos with superimposed contour lines, from government agencies. The departments of transportation, mapping, or flood control are usually a good place to start.

G (CHAPTER 2). METEOROLOGICAL AND CLIMATE RESOURCES

ONLINE RESOURCES

www.wrh.noaa.gov. This is the United States National Weather Service's website. Locate the weather stations closest to your site and find out their elevations. Download data from those stations that are most like your site.

ag.arizona.edu/azmet. Arizona Meteorological Network. Evaporation rates, prevailing winds, soil temperatures, and minimum/maximum temperatures are listed for various sites. For other states contact your local agricultural extension service for similar meteorological networks.

MISCELLANEOUS

The U.S. National Forest Service compiles data for remote weather stations, though the data is not as comprehensive nor standardized as the above two resources. However, for rural sites a Forest Service weather station may be closer to a given site than one monitored by other agencies.

Local airports, since they collect and record climatic data.

Rain gauge from a local hardware or garden store with which to begin keeping precipitation records for your site.

H (CHAPTER 3). RAINWATER HARVESTING WITH EARTHWORKS RESOURCES

PUBLICATIONS

Rainwater Harvesting for Drylands, vol. 2: Water-Harvesting Earthworks, by Brad Lancaster. Rainsource Press, www.HarvestingRainwater.com, 2006. A thorough guide describing how to create and use diverse water harvesting earthworks and their numerous variations. Many case studies are featured along with tips on how to integrate your earthworks with multiple on-site resources and challenges so they do far more than harvest water. Includes a chapter on integrating the harvest of greywater.

Drylands Watershed Restoration: Introductory Workshop Activities, by Ben Haggard. Sol y Sombra Foundation, 1994. A wonderful resource about water harvesting earthworks and how to set up workshops on the subject, although hard to find nowadays.

An Introduction to Erosion Control, by Bill Zeedyk and Jan-Willem Jansens. Earth Works Institute, Rio Puerco Management Committee, Quivira Coalition, May 2004. Basic how-to guide for simple and effective erosion control strategies that harvest soil and water.

"Dynamic Water Storage," by Tim Murphy. *Permaculture Drylands Journal*, #30, Summer 1998, pp. 22–24.

Alternative Irrigation: The Promise of Runoff Agriculture, by Christopher J. Barrow. Earthscan, 1999. Earthscan Publications Ltd., 120 Pentonville Road, London, N1 9JN, UK. An introduction to strategies of runoff agriculture used around the world.

VIDEOS/DVDS

Harvest Rain, by the Fundacion San Bernardino. Contact Valer Austin: vaustin@elcoronadoranch.net. Highlights the dramatic success of constructing check dams in the watersheds of El Coronado Ranch, Arizona

I (CHAPTER 3). RAINWATER HARVESTING WITH CISTERNS RESOURCES

PUBLICATIONS

Rainwater Harvesting for Drylands, vol. 3: Roof Catchment and Cistern Systems, by Brad Lancaster. Rainsource Press, www.HarvestingRainwater.com, 2006. This volume presents guidelines specific to cistern systems, helps you select the best non-toxic materials for your system, and size it for maximum efficiency. Numerous case studies are given describing various systems and how, through integrated design, their tanks do far more than just store water.

Rainwater Collection for the Mechanically Challenged, 2nd rev. ed., by Suzy Banks with Richard Heinichen. Tank Town Publishing, 2004. Tank Town Publishing, 1212 Quail Ridge, Dripping Springs, Texas 78620; phone: 512-894-0861; website: www.rainwatercollection.com.

An entertaining resource on various cisterns and how you could plumb them into your home with gravity-fed or mechanically pressurized systems.

Rainwater Catchment Systems for Domestic Supply: Design, Construction, and Implementation, by John Gould and Erik Nissen-Petersen. Intermediate Technology Publications, 1999. Intermediate Technology Publications, 103/105 Southhampton Row, London, WC1B 4HH, UK; website: http://www.itdgpublishing.org.uk/. An excellent review of rainwater harvesting practices around the world. It presents case studies which will help anyone intending to design or construct a rainwater catchment system.

Texas Guide to Rainwater Harvesting, 3rd ed., by Wendy Price Todd and Gail Vittori. Texas Water Development Board in Cooperation with the Center for Maximum Potential Building Systems, 1997. For a printed copy write to Conservation, Texas Water Development Board, P.O. Box 13231, Austin, Texas 78711-3231. Available online at www.twdb.state.tx.us/publications/reports/RainHarv.pdf. A great, easy to read resource on harvesting rainwater in cisterns. Water quality, basic system set up, and case studies are covered. A short video covering cistern basics is also available.

Water Storage: Tanks, Cisterns, Aquifers and Ponds for Domestic Supply, Fire and Emergency Use, Plus How to Make Ferrocement Water Tanks, by Art Ludwig. Oasis Design, www.oasisdesign.net, 2005. A great overview and how-to on numerous water storage options.

Ferrocement Water Tanks and Their Construction, by S. B. Watt. Intermediate Technology Publications, 1978. A very practical resource emphasizing low-tech methods of ferrocement cistern construction, and mention of unstabilized adobe cisterns.

Guidelines on Rainwater Catchment Systems for Hawaii, by Patricia S. H. Macomber. College of Tropical Agriculture and Human Resources, University of Hawaii at Manoa, 2001. Available online at www2.ctahr.hawaii.edu/oc/freepubs/pdf/RM-12.pdf.

An informative guide documenting the use of various tanks, and providing a good overview of harvested rainwater quality and treatment and filtration options.

Guidance on the Use of Rainwater Tanks, 2nd ed., enHealth Council of the National Public Health Partnership. Government of South Australia, 2004. ISBN 0 642 82443 6; order from website. www.dh.sa.gov.au/pehs/publications/publications.htm. Good deal of information and management tips on maintaining cistern water quality.

Sustainable House, by Michael Mobbs. Choice Books, 1998. How a family of four renovated their inner-city Sydney, Australian home to make it almost entirely self-sufficient in electricity, water, and waste disposal.

"The Secrets Of Low Tech Plumbing," by John Vivian. *Mother Earth News*, June/July 1995, pp. 34–90. Now available on the Mother Earth News website: www.motherearthnews.com/top_articles/1995_June_July/The_Secrets_of_Low_Tech_Plumbing. Information on simple rain catchment, water ram pumps, solar pumps, and old fashioned water conservation.

ONLINE RESOURCES (IN ADDITION TO ABOVE)

www.watertanks.com. A good quick reference for availability and prices of pre-manufactured water tanks. You may find a local distributor is more convenient and cheaper without the shipping, but this website is a good start.

http://www.harvestingwater.com/rainwatr.htm. This is a great website documenting Ole Ersson's permitted potable rainwater harvesting system in Portland, Oregon. Includes diagram of the system and components used. The system cost $1,500 and harvests 27,000 gallons per year.

VIDEOS

Rainwater Collection for the Mechanically Challenged. Tank Town, 1999. Tank Town, PO Box 1541,

Dripping Springs, Texas 78620; phone: 512-894-0861; website: www.rainwatercollection.com. An entertaining and informative 37-minute video on the design and installation of household rainwater systems providing potable water.

J (CHAPTER 3). HOUSEHOLD WATER USAGE AND WATER CONSERVATION

PUBLICATIONS

Handbook of Water Use and Conservation, by Amy Vickers. WaterPlowPress, 2001. Ordering: call 866-367-3300 or online at www.waterplowpress.com. The most thorough reference on water use and conservation.

Water Efficiency for Your Home: Products and Advice Which Save Water Energy, and Money, 3rd ed., by John Woodwell, Jim Dyer, Richard Pinkham, and Scott Chaplin. Rocky Mountain Institute, 1995. Available online: www.rmi.org/images/other/Water/W95-36_WaterEff4Home.pdf

ONLINE RESOURCE

www.h2ouse.org. This user-friendly website provides water use rates for household appliances, and recommended conservation strategies.

K (CHAPTER 2). GREYWATER RESOURCES

PUBLICATIONS

Greywater and Your Detergent, prepared by the Office of Arid Land Studies in cooperation with the Soil, Water and Plant Analysis Laboratory, University of Arizona, sponsored by Tucson Water (phone: 520-791-4331). A handy pamphlet comparing the performance of various detergents dissolved in greywater used to irrigate a landscape.

Create an Oasis with Greywater: Your Complete Guide to Choosing, Building and Using Greywater Systems, by Art Ludwig. Oasis Design, 1994–2000. Oasis Design, 5 San Marcos Trout Club, Santa Barbara, CA 93105-9726; website: www.oasisdesign.net. Presents a wide array of greywater-harvesting options.

Branched Drain Greywater Systems: Reliable, Economical, Sanitary Distribution of Household Greywater to Downhill Plants Without Filtration or Pumping, by Art Ludwig. Oasis Design 2000-2002. Oasis Design, see above for address; website: www.oasisdesign.net. My favorite greywater book and system.

Greywater Guidelines, by Val L. Little. To obtain a copy (hardcopy or downloadable pdf), see http://watercasa.org/pubs/graywaterguidelines.html. Arizona greywater guidelines.

ONLINE RESOURCES

www.greywater.com. Great resources for greywater, including various systems and greywater volume estimates.

www.oasisdesign.net. Art Ludwig's website, probably the most comprehensive web resource for greywater, and you'll find plenty more info relating to water and ecological design.

L (CHAPTER 4). SUN ANGLES AND PASSIVE SOLAR DESIGNS

PUBLICATIONS

The Passive Solar Energy Book, by Ed Mazria. Rodale Press, 1979. Out of print, but available from amazon.com, powells.com, and others. Excellent.

Sun, Wind, and Light: Architectural Design Strategies, 2nd ed., by G. Z. Brown and Mark DeKay. William Stout Architectural Books, 2000. A pattern book illustrating passive heating and cooling strategies for a diverse array of contexts ranging from individual homes to high rises to whole towns.

The Food and Heat Producing Solar Greenhouse, by Bill Yanda and Rick Fisher. John Muir Publications, 1980.

Effective Shading with Landscape Trees, by William B. Miller and Charles M. Sacamano. University of Arizona College of Agriculture, Cooperative Extension bulletin 188035/8835, March 1990.

OTHER

Your site's latitude: Look at a globe, atlas, or topographic map, or google "what is the latitude of (your town, state, country)."

SOLAR OVENS/COOKING

Sun Ovens International, Inc. www.sunoven.com. They make the Global Sun Oven®, the most efficient solar oven I've used.

Solar Cookers International. www.solarcookers.org. A great organization/resource for making and using your own solar oven.

M (CHAPTER 4). SOLAR RIGHTS

New Mexico's solar rights: For information contact the Energy Conservation and Management Division, NM Energy, Minerals and Natural Resources Department, 1220 S. Saint Francis, Santa Fe, NM 87505; phone: 505-476-3310; website: www.emnrd.state.nm.us/ecmd/html/solar.htm

N (CHAPTER 4). INTEGRATED DESIGN PATTERNS

A Pattern Language: Towns, Buildings, Construction, by Christopher Alexander, S. Ishikawa, M. Silverstein. Oxford, 1977.

An Introduction to Permaculture, by Bill Mollison. Tagari Publications, 1988.

Sun, Wind, and Light: Architectural Design Strategies, 2nd ed., by G. Z. Brown and Mark DeKay. William Stout Architectural Books, 2000.

The Hand-Sculpted House: A Practical and Philosophical Guide to Building a Cob Cottage, by Ianto Evans, Michael G. Smith, and Linda Smiley. Chelsea Green Publishing, 2002.

Designing and Maintaining Your Edible Landscape—Naturally, by Robert Kourik. Metamorphic Press, 1986.

The following resource sections are about rainwater-harvesting community and government resources, as well as a list of firms and individuals who do rainwater-harvesting design.

O. GREEN HOME TOURS

Tour other sites to learn from their successes and mistakes.

American Solar Energy Society National Solar Home Tour. Search the website under "tour": www.ases.org

National Association of Home Builders Green Building Program. Search the website under "green building program": www.nahbrc.org. This program can put you in contact with local green builder programs, some of which organize tours.

P. MODELS FOR PUBLIC SITES

Akash Ganga Chennai Rain Centre. http://akash-ganga-rwh.com/RWH/WaterHarvesting.html. An urban rainwater harvesting demonstration site in Chennai, India.

"Rainwater Harvesting: Success Story from Chennai India," report by Ram Krishnan presented at the ARCSA Conference in Austin, Texas, August 21-23, 2003. Order proceedings from www.arcsa-usa.org.

TreePeople website: www.TreePeople.org/trees/.
Features water-harvesting demonstration sites in Los
Angeles, California.

Seattle, Washington Public Utilities
SEA Streets Project. Website:
http://www.ci.seattle.wa.us/util/About_SPU/Drainage_&
_Sewer_System/Natural_Drainage_Systems/index.asp.
Progressive multi-use water harvesting/beautification/flood
control strategies in the public rights-of-way.

Q. SUSTAINABLE COMMUNITIES

Superbia!: 31 Ways to Create Sustainable Neighborhoods,
by Dan Chiras and Dave Wann. New Society
Publishers, 2003. Presents a number of easy ways to
improve existing communities.

*Designing Sustainable Communities: Learning from
Village Homes*, by Judy Corbett and Michael Corbett.
Island Press, 2000. An inside look at the development
of Village Homes, a unique example of sustainable
design in a community. This 60-acre residential and
business development includes extensive common
areas, community gardens, narrow streets, pedestrian
and bike paths, solar homes, and in terms of rainwater
harvesting—an innovative ecological drainage system.

R. WATER-HARVESTING RESOURCES LIST

Travelers Earth Repair Network.
www.geocities.com/Rainforest/4663/tern.html.
A project of the Friends of the Trees Society, with
links to many other resources.

S. WATER-HARVESTING GROUPS TO JOIN

American Rainwater Catchment Systems Association:
www.arcsa-usa.org.

International Rainwater Catchment Systems
Association: www.ircsa.org.

T. GROUPS OFFERING WORKSHOPS IN RAINWATER HARVESTING AND PERMACULTURE, INFORMATION SOURCES

MAGAZINES/JOURNALS

Permaculture Activist. Information on courses and
events, as well as subscription information at
www.permacultureactivist.net. Permaculture and
related courses around the world.

*The Last Straw: The International Journal of Straw Bale
and Natural Building*. www.thelaststraw.org. Lists natural building and sustainable building events and
workshops.

ORGANIZATIONS

Drylands specific

Sonoran Permaculture Teaching Guild. Website:
www.sonoranpermaculture.org

Permaculture Institute, USA, Casa Las Barrancas
Farm, PO Box 3702, Pojoaque, NM 87501, USA.
Phone/Fax: 505-455-0270.
Email: pci@permaculture-inst.org

Ecoversity, 2639 Agua Fria, Santa Fe, NM 87505.
Phone: 505-424-9797. Website: www.ecoversity.org

Earthworks Institute, 1413 Second St., Suite 4, Santa
Fe, NM 87505. Phone: 505-982-9806. Website:
www.earthworksinstitute.org

Non-Drylands Specific

Occidental Arts And Ecology Center (OAEC) and
OAEC's Water Institute, 15290 Coleman Valley Road,
Occidental, CA 95465. Phone: 707-874-1557.
Website: www.oaec.org. OAEC and the WATER
Institute conduct many fine workshops including
"Watershed Basins of Relations: Starting and
Sustaining Community Watershed Groups."

Permaculture Institute of Northern California, PO Box 341, Point Reyes Station, CA 94956. Phone: 415-663-9090. Website: www.permacultureinstitute.com

In Mexico

Alternativas, Tehuacan, Puebla, MEXICO. Website: www.alternativas.org.mx. Email: info@alternativas.org.mx

U. WATERSHED COMMUNITY ORGANIZING AND WATERSHED AWARENESS

PUBLICATIONS

"Basins of Relations: Restoring a Watershed State of Being," by Brock Dolman. *Permaculture Activist*, no. 47, Summer 2002, pp. 8–12.

"A Watershed Runs through You," by Freeman House. *YES! Magazine*, no. 28, Winter 2004.

Watersheds: A Practical Handbook for Healthy Water, by Clive Dobson and Gregor Gilpin Beck. Firefly Books, 1999. A beautifully illustrated book providing an overview of the fundamentals of ecology from the simple concept of a watershed to the biological intricacies of a wetland ecosystem and its implications on the environment.

Getting in Step: Engaging and Involving Stakeholders in Your Watershed, by Charlie MacPherson, Barry Tonning, and Emily Fallasli of Tetra Tech, Inc. U.S. Environmental Protection Agency, 68-C-99-249, 1998. Contact Charlie MacPherson at 703-385-6000 or email Charlie.macpherson@tetratech-ffx.com to get a copy. This guide provides the tools needed to effectively engage stakeholders to restore and maintain healthy environmental conditions throughout their watershed through community support and cooperative action.

Getting in Step: A Guide for Conducting Watershed Outreach Campaigns, prepared by Tetra Tech, Inc. U.S. Environmental Protection Agency, EPA 841-B-03-002, December 2003. For copies of this guide and its companion video contact: National Service Center for Environmental Publications, Phone 800-490-9198; website www.epa.gov/ncepihom. This guide is an update of the 1998 publication "Getting in Step: A Guide to Effective Outreach in Your Watershed." This updated version includes more specific information on how to work with the mass media to conduct an outreach campaign.

Starting Up: A Handbook for New River and Watershed Organizations, compiled by Katherine Luscher. River Network, 1996. For copies visit www.rivernetwork.org or phone 503-241-3506. This 440-page handbook is based on the experience of dozens of veteran leaders in the river and conservation movements with articles laying out the critical moves every newly forming organization needs to thrive and grow.

How to Start a Watershed Awareness Program, by the Aquatic Outreach Institute. Available from the Watershed Project Store; phone 510-231-5655; website: www.thewatershedproject.org

Stormwater Strategies: Community Responses to Runoff Pollution, by Peter H. Lehner, George P. Aponte Clarke, Diane M. Cameron, and Andrew G. Frank. Natural Resources Defense Council (NRDC), May 1999. One hundred case studies of successful projects around the U.S. that simultaneously improve runoff quality and the environment, have economic advantages, and additional community benefits.

"Stormwater Management: Use It or Lose It," by Tim Murphy. *Sustainable Living in Drylands*, no. 5, Winter 1988/89. A great wake-up call to the value of our stormwater runoff, and how we can use it as the local resource it is.

Stormwater: Asset Not Liability, published by the Los Angeles and San Gabriel Rivers Watershed Council. Phone: 213-367-4111.

Stormwater Journal keeps you updated on issues related to stormwater control and lessening stormwater runoff pollution. Subscription information: www.stormh2o.com.

International Erosion Control Association (IECA) and its publication the *Erosion Control Journal* keep you updated on erosion control strategies pushed by regulators and the industry. Membership and other information: www.ieca.org.

PROGRAMS

Occidental Arts and Ecology Center's WATER Institute, 15290 Coleman Valley Road, Occidental, CA 95465. Phone: 707-874-1557. Website: www.oaec.org. This organization offers the four-day residential training program, "Basins of Relations: Starting and Sustaining Community Watershed Groups."

V. DESIGNING URBAN LANDSCAPES AND RETROFITTING CITIES AS A SERIES OF FUNCTIONING MINIATURE URBAN-FOREST WATERSHEDS

PUBLICATIONS

Second Nature: Adapting LA's Landscape for Sustainable Living, edited by Patrick Condon and Stacy Moriarty. Metropolitan Water District of Southern California, 1999. A great resource of a group in Los Angeles creating cross-jurisdictional and cross-disciplinary connections between the people and institutions responsible for the infrastructure, planning, and ecology of Los Angeles in order to view the city as a living watershed. Concepts such as passive rainwater harvesting and multiple-use landscaping are presented that could help improve the sustainability of the City and the watershed. Contact the organization at: TreePeople, 12601 Mulholland Drive, Beverly Hills, CA 90210; website: www.TreePeople.org/trees.

Product Specification for the Transagency Resources for Economic and Environmental Sustainability Project, by John Stokes Associates, Inc., 1998. Prepared for TreePeople, Beverly Hills, CA. This is the Cost-Benefit Analysis for the T.R.E.E.S. Project, a program in Los Angeles. See *Second Nature*, above.

ONLINE RESOURCES

TreePeople website: www.TreePeople.org/trees/

W. SOURCE OF LOANS FOR WATER-HARVESTING PROJECTS

Permaculture Credit Union. Office location: 4250 Cerrillos Road, 2nd floor, Santa Fe, NM 87507. Mailing Address: PO BOX 29300, Santa Fe NM 87592-9300. Phone: 866-954-3479 or 505-954-3479. Fax: 505-424-1624. Website: www.puconline.org. Email: perma@pcuonline.org

X. FIRMS/DESIGNERS SPECIALIZING IN DRYLAND WATER HARVESTING, PERMACULTURE DESIGN, OR INTEGRATED DESIGN

This is only a partial list of designers and implementation firms doing such work. Those listed helped me with this book through their review of the text and sharing their work and knowledge.

Santa Fe Permaculture, 551 W. Cordova Road #458, Santa Fe, NM 87505. Phone 505-424-4444. Website: www.sfpermaculture.com.

Regenesis Group. Ecological resources for communities, designers, and development professionals. Website: www.regenesisgroup.com.

San Isidro Permaculture, 1517 Camino Sierra Vista, Santa Fe, NM 85705. Phone: 505-983-3841, 505-501-GROW. Email: sanisidroperm@hotmail.com.

Earthwrights Designs, 30 Camino Sudeste, Santa Fe, NM 87508. Phone: 505-986-1719. Email: ezentrix@aol.com.

Rocky Brittain, architect. Phone: 520-884-8226.

Heather Kinkade-Levario. Phoenix, Arizona. Website: www.forgottenrain.com.

David Confer, integrated systems consultant. Phone: 520-991-3737.

David Omick, sustainable systems/appropriate technology designer. Website: www.omick.net.

Art Ludwig, of Oasis Designs, greywater and sustainable systems, designer/author. Website: www.OasisDesign.net.

Peter Pfieffer, of Barley and Pfeiffer Architects, 1800 W. 6th St., Austin, Texas 78703. Phone 512-476-8580. Website: www.barleypfeiffer.com. Designs swimming-pool-style below-ground sprayed-concrete tanks.

Overland Partners/Architects, 5101 Broadway, San Antonio, Texas 78209. Website: www.overlandpartners.com. Designed the water-harvesting system for the Wildflower Center in Austin, Texas.

Steve Kemble, of Sustainable Systems Support, 3063 N. Camino de Oeste, Tucson, AZ 85745. 520-743-3828. Spearheaded a student-built ferrocement cistern project for East Campus high school in Douglas, Arizona.

References

INTRODUCTION

1. Alduenda, Eileen, ed., Sustainable Design: A Planbook for Sonoran Desert Dwellings (Tucson Institute For Sustainable Communities, 1999).

2. United Nations Environment Programme (UNEP), *World Atlas of Desertification*, 2nd ed. (London: Arnold; New York: John Wiley & Sons, 1997).

3. Ibid.

4. Todd, Wendy Price and Gail Vittori, *Texas Guide to Rainwater Harvesting* (Austin: Texas Water Development Board in Cooperation with the Center for Maximum Potential Building Systems, 1997).

5. Personal communication in an interview with Brian Barbaris, Senior Research Specialist, Department of Atmospheric Sciences, University of Arizona, 12 February 2003.

6. Todd and Vittori, *Texas Guide*.

7. Ibid.

8. Begeman, John, "Thanks to Storms, Rain Delivers More Than Water to Desert," *Arizona Daily Star*, Aug. 2, 1998, Home Section, p. 1.

9. Cleveland, David and Daniela Soleri, *Food From Dryland Gardens* (Center for People, Food and Environment, 1991).

10. Ibid.

11. Ibid.

12. Evenari, Michael, Leslie Shanan, and Naphtali Tadmor, *The Negev: The Challenge of the Desert* (Cambridge: Harvard University Press, 1971).

13. Gould, John and Erik Nissen-Petersen, *Rainwater Catchment Systems for Domestic Supply: Design, Construction, and Implementation* (London: Intermediate Technology Publications, 1999).

14. Bowden, Charles, *Killing the Hidden Waters* (Austin: University of Texas Press, 1977), 119–120.

15. Ibid.

16. Barlow, Maude, *Blue Gold: The Global Water Crisis and the Commodification of the World's Water Supply*. A Special Report issued by the International Forum on Globalization, June 1999.

17. Ibid.

18. Shogren, Elizabeth, "Sprawl Adds to Drought, Study Says," *Los Angeles Times*, August 29, 2002, p. A12.

19. American Rivers, Natural Resources Defense Council, Smart Growth America, *Report: Paving Our Way to Water Shortages: How Sprawl Aggravates the Effects of Drought*, August 28, 2002. Available online at www.smartgrowthamerica.org/Sprawl%20Report-FINAL.pdf

20. Personal communication via email correspondence with Frank Sousa, Tucson Department of Transportation and Engineering Division, Stormwater Section, 7 November 2002.

21. Condon, P. and S. Moriarty, eds., *Second Nature: Adapting LA's Landscape for Sustainable Living* (Los Angeles: Metropolitan Water District of Southern California, 1999).

22. American Rivers, *Report: Paving*.

23. Agarwal, Anil, Sunita Narain, and Indira Khurana, *Making Water Everybody's Business* (New Delhi: Centre for Science and Environment, 2001).

24. Glennon, Robert, *Water Follies: Groundwater Pumping and the Fate of America's Fresh Waters* (Washington, DC: Island Press, 2002).

25. Barlow, *Blue Gold*.

26. Brown, Lester R. and Brian Halweil, "China's Water Shortage Could Shake World Food Security," *Worldwatch*, July/August 1998.

27. Hawken, Paul, Armory Lovins, and L. Hunter Lovins, *Natural Capitalism: Creating the Next Industrial Revolution* (New York: Little, Brown and Company, 1999).

28. Barlow, *Blue Gold.*

29. Parfit, Michael, "Sharing the Wealth of Water," *National Geographic Special Edition: Water, the Power, Promise, and Turmoil of North America's Fresh Water,* 1993, p. 28.

30. Penn State, College of Education, *Investigations: Lesson 13: Lifestyles and Global Warming—Any Connection?* www.ed.psu.edu/ci/Papers/STS/gac-3/in13.htm.

31. Email correspondence with Tom Hansen of Tucson Electric Power on April 30, 2003 in which he stated "the Springerville Generating Station produces electricity with an annual average water consumption of about 0.45 kWh." Additional information can be found at www.powerscorecard.org.

32. H2ouse.org, "Save Water, Money, Energy Now!" www.h2ouse.org/action/top5.cfm

33. Woodwell, John, Jim Dyer, Richard Pinkham, and Scott Chaplin, *Water Efficiency for Your Home: Products and Advice Which Save Water, Energy, and Money,* 3rd ed. (Snowmass CO: Rocky Mountain Institute, 1995). www.rmi.org/images/other/Water/W95-36_WaterEff4Home.pdf

34. Ibid.

35. H2ouse.org, "Save Water."

36. Karpiscak, Martin M., Thomas M. Babcock, Glenn W. France, Jeffrey Zauderer, Susan B. Hopf, and Kenneth E. Foster, *Evaporative Cooler Water Use Within the City of Phoenix:: Final Report,* Arizona Department of Water Resources, Phoenix Active Management Area, April 1995.

37. Arizona Department of Water Resources, "Outdoor Water Use" pamphlet.

38. Figures determined from calculations from Pima County, Arizona, Cooperative Extension, Water Resources Center, Low 4 Program. "How to Develop a Drip Irrigation Schedule" handout.

39. Heede, Richard and Staff of Rocky Mountain Institute, *HOMEmade Money,* Rocky Mountain Institute, 1995.

40. Arizona Department of Water Resources, "Outdoor Water Use."

41. H2ouse.org, "Pool and Spa Water Savings." www.h2ouse.org

42. Corbett, Michael and Judy Corbett, *Designing Sustainable Communities: Learning from Village Homes* (Washington, DC: Island Press, 2000).

43. Ellis, William S., "The Mississippi: River Under Siege," *National Geographic Special Edition: Water, the Power, Promise, and Turmoil of North America's Fresh Water,* 1993.

44. Barlow, *Blue Gold.*

45. National Wild and Scenic Rivers System, "River and Water Facts." www.nps.gov/rivers/waterfacts.html

46. Barlow, *Blue Gold.*

47. Vickers, Amy, *Handbook of Water Use and Conservation* (Amherst MA: WaterPlow Press, 2001).

48. Ibid.

49. New Internationalist, "Factfile on Water," *New Internationalist,* April 2000.

50. Condon and Moriarty, *Second Nature.*

51. *New York Times* Special Supplement: *Water, Pushing the Limits of an Irreplaceable Resource,* December 8, 1998.

52. Barlow, *Blue Gold.*

53. Ibid.

54. Ibid.

55. Karpiscak et al., *Evaporative Cooler Water Use.*

56. Pima County, Arizona, Cooperative Extension, Water Resources Center, Low 4 Program. "How to Develop a Drip Irrigation Schedule" handout.

57. Heede et al., *HOMEmade Money.*

58. Shiva, Vandana, *Water War: Privatization, Pollution, and Profit* (Cambridge MA: South End Press, 2002).

59. United Nations Committee on Economic, Cultural and Social Rights 2002, General Comment No. 15. The right to water (articles 11 and 12 of the International Covenant on Economic, Social and Cultural Rights). Twenty-ninth Session, Geneva E/C. 12-2002-11.

60. Gelt, Joe, Jim Henderson, Kenneth Seasholes, Barbara Tellman, Gary Woodard, Kyle Carpenter, Chris Hudson, and Souad Sherif, *Water in the Tucson Area: Seeking Sustainability,* Water Resources Research Center, Issue Paper #20, Summer 1999.

61. Laney, Nancy, *Desert Water: From Ancient Aquifers to Modern Demands* (Tucson: Arizona-Sonora Desert Museum Press, 1998).

62. Logan, Michael F., *The Lessening Stream: An Environmental History of the Santa Cruz River* (Tucson: University of Arizona Press, 2002).

63. Ibid.

64. Ibid.

65. Arizona Department of Environmental Quality, *Superfund Programs Section Site and Program Information—C 02-04,* July 2002.

66. Laney, *Desert Water.*

67. *Science Daily,* "Possible Climate Shift Could Worsen Water Deficit in the Southwest", 2000-02-16. www.sciencedaily.com/releases/2000/02/000216052551.htm.

68. American Rivers, "Colorado River 'Most Endangered'," April 14, 2004 press release. www.americanrivers.org/site/News2?abbr=AMR_&page=NewsArticle&id=6699

69. Bergman, Charles, *Red Delta: Fighting for Life at the End of the Colorado River* (Golden CO: Fulcrum, 2002).

70. Reisner, Marc, *Cadillac Desert* (New York: Penguin Books, 1986).

71. Personal communication with Dr. Jim Riley, Associate Professor Soil, Water and Environmental Science Department, University of Arizona, 16 June 2005.

72. Vincent, Kirk and Laurie Wirt, "Urban Runoff—Lessen the Strain on Public Works by Using That Water at Home," *The Arizona Hydrological Society Newsletter,* v. 10, 1993, pp. 1–3.

73. Personal email correspondence with David Confer on 16 February 2002.
74. Ibid.
75. Personal communication with Frank Ramberg, Research Scientist, Department of Entomology, University of Arizona on 24 January 2005.

CHAPTER 1

1. *PELUM Association*, "Water Harvesting: Some General Principles and Methods for Areas of Intensive Use and Dryland Cropping." PELUM Association, Box CY301, Causeway, Harare, Zimbabwe, July 1995.
2. Ibid.
3. Agarwal, Anil, Sunita Narain, and Indira Khurana, *Making Water Everybody's Business* (New Delhi: Centre for Science and Environment, 2001).
4. Ibid.
5. Ibid.
6. Ibid.
7. *PELUM Association*, "Water Harvesting."
8. Ibid.
9. Mollison, Bill, *Introduction to Permaculture* (Tyalgum, Tasmania: Tagari Publications, 1988).
10. Ibid.
11. Ibid.
12. Ibid.
13. Ibid.

CHAPTER 2

1. Haggard, Ben, *Drylands Watershed Restoration: Introductory Workshop Activities*, Sol y Sombra Foundation. (Santa Fe: Center for the Study of Community, 1994).
2. Ibid.
3. U.S. Environmental Protection Agency (EPA), "What is a Watershed?" www.epa.gov/win/what.html
4. Figures determined from calculations from Pima County, Arizona, Cooperative Extension, Water Resources Center, Low 4 Program. "How to Develop a Drip Irrigation Schedule" handout.
5. Ludwig, Art, *Branched Drain Greywater Systems* (Santa Barbara CA: Oasis Design, May 2000–March 2002).

CHAPTER 3

1. Tipton, Jimmy L., *Water Requirements of Landscape Trees: Final Report* (Phoenix, Arizona Department of Water Resources, 1997).

CHAPTER 4

1. Phillips, Ann, Editor. *City of Tucson Water Harvesting Guidance Manual*, City of Tucson, Department of Transportation, Stormwater Section, June 2003.

2. Hammond, Jonathan, Marshall Hunt, Richard Cramer, and Loren Neubauer, *A Strategy for Energy Conservation: Proposed Energy Conservation and Solar Utilization Ordinance for the City of Davis, California*, City of Davis, August 1974.
3. Mazria, Ed, *The Passive Solar Energy Book: A Complete Guide to Passive Solar Home, Greenhouse and Building Design.* (Emmaus PA: Rodale Press, 1979).
4. Karpiscak, Martin M., Thomas M. Babcock, Glenn W. France, Jeffrey Zauderer, Susan B. Hopf, and Kenneth E. Foster, *Evaporative Cooler Water Use Within the City of Phoenix: Final Report*, Arizona Department of Water Resources, Phoenix Active Management Area, April 1995.
5. Plant water use figures determined from calculations from Pima County, Arizona, Cooperative Extension, Water Resources Center, Low 4 Program. "How to Develop a Drip Irrigation Schedule" handout.
6. Heede, Richard and Staff of Rocky Mountain Institute, *HOMEmade Money*, Rocky Mountain Institute, 1995.
7. Ibid.
8. Rosenow, John, "Every Tree Is Part of the Global Forest," *Arizona Urban and Community Forestry*, vol. 5, No. 1 (March 1999), p. 3.
9. Ibid.
10. Corbett, Judy and Michael Corbett, *Designing Sustainable Communities: Learning from Village Homes* (Washington, DC: Island Press, 2000).
11. Kourik, Robert, *Designing and Maintaining Your Edible Landscape Naturally* (Santa Rosa, CA: Metamorphic Press, 1986).
12. Corbett and Corbett, *Designing Sustainable Communities*.
13. Milagro Co-Housing Declaration of Covenants, Conditions, and Restrictions.
14. "Recognizing Skin Cancer," Boston University, 1997. www.bu.edu/cme/modules/2002/skincancer02/content/04-malig.html
15. Mobely, Barbara, "Extension Launches Skin Cancer Inititiative," Alabama Cooperative Extension System. www.aces.edu/dept/extcomm/newspaper/may20a03.html
16. Keating, Janis, "TREES: The Oldest New Thing in Stormwater Treatment?" *Stormwater*, March 2002. www.forester.net/sw_0203_trees.html
17. Hammond et al., *Strategy for Energy Conservation*.
18. Corbett and Corbett, *Designing Sustainable Communities*.
19. Ibid.
20. James, William, "Green Roads: Research Into Permeable Pavers," *Stormwater*, March 2002. www.forester.net/sw_0203_green.html

Glossary

Aerobic. A condition that supports organisms that only exist in the presence of oxygen.

Algae. Microscopic plants which contain chlorophyll and live in water. Algae can impart tastes and odors to stored water.

Anaerobic. A condition that supports organisms that only exist in the absence of oxygen.

Angular open-graded gravel. Aggregate whose particles are angular in shape so their flat faces interlock with each other to resist rotating and shifting, and having a narrow range of particle sizes and open void spaces that improve interlocking between the particles, while maintaining good porosity.

Annual. Plant that takes 1 year or less to go through its entire life cycle: germination of the seed, vegetative growth, flowering, and seed production, after which it dies.

Aquifer. Subterranean layers of sedimentary particles (sand, gravel, and rocks) laid down over geologic time, in which water fills the tiny spaces between the particles.

Backwater valve. A sewer drain pipe valve consisting of a flap that opens with water flowing out, but that otherwise remains closed to prevent the inflow of backed-up sewage. It is commonly made from ABS plastic, typically used with gravity-fed sewer drains, and is available from better-stocked plumbing suppliers. I install backwater valves at the ends of cistern overflow pipes to keep insects, critters, and sunlight out of the stored water.

Berm 'n basin. A water-harvesting earthwork laid perpendicular to land slope, consisting of an excavated basin and a raised berm located just downslope of the basin.

Biocompatible. A material whose breakdown materials produced during decomposition are beneficial for, or at least not harmful to, the environment in which it is disposed.

Blackwater. Wastewater from toilets (some regulators consider kitchen sink wastewater as blackwater too), that has higher levels of solids and coliform bacteria than greywater sources.

Boomerang berm. Semicircular berm open to, and harvesting, incoming runoff from upslope.

Branched drain greywater system. System of pipes, valves, or "double L" or "Y" fittings that "branch" or split a gravity-fed flow of greywater to as many as sixteen outlets within mulched basins distributed in a landscape.

Catchment surface. Surface from which runoff is captured within earthworks or a cistern for beneficial on-site use.

Central Arizona Project (CAP). A multi-billion dollar canal project which diverts water from the Colorado River and pumps it 1,000 feet uphill and over 300 miles (482 km) through the desert to reach farms and the cities of Phoenix and Tucson, Arizona.

Channel flow. The concentrated distribution of runoff within distinct channels or drainages. Look for nick points, rills, gullies, bank cutting, different sediment sizes, vegetation growing within channels, and exposed roots to assess the force of the flow and the health of the channel.

Channelization. Constricting and straightening water flow by sealing and smoothing the banks and sometimes the bed of a waterway, often with concrete. It can be compared to a shotgun barrel for water. Channelization increases the velocity of water flow through and downstream of the channelized area, reducing infiltration of water into the soil and sometimes deepening the channel.

Check dam. A low, leaky barrier placed perpendicular to the flow of water within a drainage to slow the water's flow, infiltrate more water into the soil, and hold soil and organic matter higher in the watershed.

Cistern. A tank used to store rainwater.

Combined sewer. Sewer that contains sewage, household wastewater, and rain runoff from streets, yards, and driveways.

Commodify. To turn a natural resource into a limited-access commodity to be bought, sold, and hoarded.

Commons. A natural resource or ecosystem that provides the ecological basis of life and whose sustainability and equitable allocation depends on cooperation among its community members.

Communify. To work together to enhance a natural resource and the related community by managing the sustainable, fair use and equal accessibility of the resource.

Community. Represents all living and interacting organisms in an ecosystem, including people, other animals, plants, fungi, and bacteria.

Compost. A soil amendment made from decomposed organic matter. The act of composting speeds up the decomposition of the organic matter, while retaining more nutrients by keeping the compost pile moist (in a pit, in the shade, covered in mulch), lightly aerated, and by balancing the amount of carbon material (dry woody material like straw or sawdust) to nitrogen-rich material (green plant material, fresh manure, urine), which also prevents odors.

Composting toilet. A waterless toilet in which dry carbon-rich material such as straw or sawdust is added to its aerobic composting chamber to help decompose (without objectionable odor) nitrogen-rich human feces and urine into high-grade fertilizer.

Contour berm. A berm 'n basin constructed along a contour line.

Contour line. A level line perpendicular to the slope of the land.

Culvert. A drainage pipe made to transport water beneath a roadway. Metal culverts can be made into above-ground cisterns.

Daylighting pipe. Outletting a pipe into the open air.

Daylighting a waterway. Uncovering and revegetating a previously piped or buried waterway to recreate a natural, living watercourse.

Degenerative. An investment that starts to degrade or break down as soon as it is made, requires on-going investments of energy and outside inputs to keep it functional, consumes more resources than it produces, and typically serves only one function.

Detention/Retention basin. A structure that decreases stormwater flow from a site by temporarily holding the runoff on site. This is not a water-harvesting structure unless the held water is beneficially utilized on site (irrigating vegetation for example).

Diversion swale. A gently sloping drainageway that moves water slowly downslope across a landscape, while simultaneously allowing some of it to infiltrate into the soil.

Drip irrigation. An irrigation strategy applying water via an emitter to the root zone of a plant at a rate slow enough (usually less than 3 gallons (12 liters) per hour) to allow the soil to absorb it without runoff.

Dryland. Areas of the world where potential average yearly moisture loss (evapotranspiration) exceeds average yearly moisture gain (precipitation).

Dry-stacked retaining wall. A naturally porous wall of stone, brick, or salvaged concrete laid "dry" without mortar, and maintaining a batter or lean of 5 to 15° into the slope to help counter the weight of upslope earth.

Dry system downspout. A dry system downspout drains all runoff water directly out of the pipe, leaving the downspout dry between rainstorms. A dry system downspout does not collect sludge, nor is it prone to freezing damage.

Dumpster. An exterior trash container from which opportunistic scavengers can salvage discarded resources.

Ephemeral water flow. Water that only flows seasonally or during and just after storms.

Erosion. Wearing away of soil and rock by gravity, wind, and water, intensified by human land-clearing practices.

Evaporation. The change of water from a liquid to a gas.

Evapotranspiration. The combined measurement of water loss to evaporation and transpiration through the pores of vegetation.

Ferrocement. Metal-reinforced cement mortar.

First flush system. A device or length of capped pipe that diverts the dirtiest or foulest first flush of water running off a catchment surface away from a cistern.

French drain. A trench or basin filled with porous materials such as gravel or mulch that have ample air spaces between them allowing water to infiltrate quickly into the drain and percolate into the root zone of the surrounding soil, while creating a stable surface you can walk on.

Gabion. A check dam in which the rocks are encased in a wrapping of wire fencing or a wire basket that holds everything together—sort of a rock burrito in a wire tortilla.

Gabion basket. A rectangular wire basket made to contain many rocks, forming a check dam across a drainage.

Generative. An investment that starts to degrade as soon as it is made, requires on-going investments of energy and outside inputs to keep it functional, produces more resources than it consumes, and typically serves multiple functions.

Greywater. Wastewater originating from a clothes washer, bathtub, shower, or sink that can be safely reused to irrigate a landscape.

Greywater harvesting. The practice of safely directing the greywater generated at a site to the root zone of perennial plants in the yard where it can help grow beautiful and productive landscapes.

Greywater stub out. A greywater plumbing connection installed during a building's construction or remodeling, allowing easy and inexpensive future access to the drainwater stream. To utilize the greywater, a simple greywater distribution system is set up within the landscape, then hooked up to the stub out.

Groundwater. Water that has naturally infiltrated into and is stored within an underground aquifer.

Guild. A harmonious assembly of living species such as plants, animals, and people and non-living elements such as rocks or buildings that perform better through their cooperative interrelationships than they would as individuals.

Gully. A large erosive drainage or arroyo.

Hard water. A characteristic of water containing dissolved calcium and magnesium, which is responsible for most scale formation in water heaters and pipes.

Hardscape. Hard paving material such as concrete sidewalks, asphalt streets, and brick patios.

Headcut. The growing upstream edge of an erosive gully or rill.

Hydrologic cycle. The continual movement of water between the earth and the atmosphere through precipitation, infiltration into and release from living systems, evaporation, evapotranspiration, and precipitation again.

Impervious. A non-permeable solid surface.

Imprinter roller. An imprinting tool fabricated from a 10 to 20-foot (300-600 cm) long, 20 or 24-inch (50–60 cm) diameter smooth roller with 10-inch (25 cm) lengths of 6 X 6-inch (15 cm) to 8 X 8-inch (20 cm) angle iron welded onto the roller in a pattern of staggered star rings.

Imprinting. A water-harvesting technique used to accelerate the revegetation of disturbed or denuded land with annual precipitation from 3 to 14 inches (76 mm to 330 mm) by creating numerous small, well-formed depressions in the soil that collect seed, rainwater, sediment, and plant litter, and provide sheltered microclimates for germinating seed and establishing seedlings.

Infiltration. The movement of water from the land's surface into the soil.

Infiltration basin. A landscaped level-bottomed, relatively shallow depression dug into the earth that collects, infiltrates, and utilizes the rain that falls within it, the runoff draining into it from the surrounding area, and potentially household greywater too.

Infiltration chamber. An empty, bottomless subsurface plastic chamber into which greywater is released, reducing direct human or animal contact with the greywater, and reducing the risk of roots growing into the greywater pipe.

Integrated design. A very efficient design methodology that provides on-site needs (e.g., water, shelter, food, aesthetics) with on-site elements (e.g., stormwater runoff, greywater, cooling shade, warming sun, vegetation) by assessing all on-site resources, and placing and designing all new elements so they build on these existing resources and help divert, deflect, or convert the site's challenges into still more resources. Integrated design saves resources (e.g., energy, water, money), while enhancing the function and sustainability of a site.

Jandy valve. My preferred, fully adjustable, three-way diversion valve for a household greywater distribution system. Available from pool and spa suppliers.

Land subsidence. The settling or sinking of land resulting from the compaction of the sedimentary layers of an aquifer when groundwater is withdrawn from the pore spaces of these sedimentary layers faster than the water can be naturally replaced.

Low-water-use vegetation. Vegetation that can subsist on natural rainfall alone.

Microclimate. A more temperate or extreme localized climate created by the shelter or exposure of adjacent landscape features or buildings.

Microorganisms. Plants or animals of microscopic size.

Mulch. A porous layer of organic matter or rock on the surface of the soil (not mixed into the soil) that increases the porosity and fertility of the underlying soil, while reducing soil moisture loss to evaporation.

Municipal water use. Use of "city water" that you pay for, and which often is piped and pumped long distances. The source is typically surface water (such as from reservoirs or rivers) or groundwater pumped from aquifers.

Native vegetation. Vegetation indigenous to a 25-mile (40-km) radius of a site and found within 500 feet (150 m) of the site's elevation. Some sites may require defining native with a larger radius to bring in more plant diversity, but the smaller the radius the more likely the plants can thrive within the climatic constraints of the site.

Natural recharge. The rate at which water naturally fills or replenishes an aquifer.

Net and pan system. A modified series of boomerang berms connected directly to one another, concentrating harvested runoff at multiple points in the landscape. A completed system looks like a "net" of berms draped over a hillside with "pans" or basins inside each segment of the "net."

Nonpoint source pollution. Pollutants from many diffuse sources. Nonpoint-source pollution is caused by stormwater or snowmelt moving over the ground. The runoff picks up and carries away natural and human-made pollutants, finally depositing them into lakes, rivers, wetlands, coastal waters, and even underground sources of drinking water.

Nonpotable water. Water that is not safe for human consumption without adequate filtration and/or treatment.

Oasis zone. The area around gathering spots like patios, front porches, and paths within 30 feet (9 m) of a home where water resources such as roof runoff and household greywater are readily available to support a greater density and/or diversity of vegetation.

On-site water budget. The amount of rainwater falling on, and runoff running onto a site, minus the amount of water lost to runoff. Sometimes greywater generated and reused on site is also included, but unlike the rainwater, this is not necessarily a sustainable water source if pumped or trucked in from off site.

Open-pollinated. Non-hybrid plants produced by transferring pollen from two parents from the same variety, which in turn produce offspring just like the parent plants. Heirloom vegetables are open-pollinated varieties passed down from generation to generation.

Organic. Non-genetically modified life grown or raised without synthetic fertilizers, pesticides, or sewer sludge in such a way that the fertility of associated soil improves with time.

Organic groundcover. Natural materials that break down and improve the soil such as dead plant material, manure, or compost.

Orientation. How a building or plantings are oriented in relationship to the winter's noonday sun, and the angles of the rising and setting sun year round. Buildings with their longer wall and windows facing the winter noonday sun, and shorter walls and fewer windows facing the rising and setting sun are much easier to passively heat and cool than a building of the opposite orientation.

Outlet chamber. An empty, bottomless subsurface chamber into which greywater is released, reducing the potential that people or animals come into direct contact with the greywater, and reducing the risk of roots growing into the greywater pipe.

Overflow. The planned and stabilized exit route for excess water from a water-harvesting earthwork or tank.

Overflow water. Excess water exceeding the storage capacity of a water-harvesting earthwork or tank.

P-trap. Drain pipe in the shape of the letter P used to prevent sewer gas from entering a building by keeping a water seal in the bend of the pipe.

Parts per million (ppm) and parts per billion (ppb). Used to quantify amounts of pollutants in water, etc.

Pathogen. An organism that may cause disease.

Peak surge. The highest short-term volume of expected water flow.

Percolation. The downward movement of water infiltrating the soil.

Perennial. Plant that lives longer than two years.

Perennial water flow. Water that continually flows year round, year after year.

Permaculture. A methodology of integrated, sustainable design based on natural systems.

Permeable paving. A broad term for water-harvesting techniques that use porous hardscape/paving materials to enable water to pass through the pavement and infiltrate into soil, passively irrigating adjoining plantings, dissipating the heat of the sun, reducing soil compaction, allowing tree roots beneath the paving to breathe, filtering pollutants, and decreasing the need for expensive drainage infrastructure.

pH. The measure of acidity or alkalinity ranging from 1 to 14. Below 7 is increasingly acid, 7 is neutral, and above 7 is increasingly alkaline.

Potable water. Water that is safe for human consumption and can be used for the greatest variety of uses.

Rainhead. An angled downspout screen placed below a roof gutter. The rainhead is designed to direct most debris off the screen to the ground below, while allowing water through and down the downspout.

Ramada. An outdoor shade structure under which it is comfortable to gather.

Regenerative. An investment that starts to grow or improve once it is made, does not require on-going investments of energy and outside inputs to keep it functional, produces more resources than it consumes, typically serves multiple functions, and can reproduce itself.

Renewable. A resource that can be replaced in a short period of time. Renewable does not necessarily mean sustainable, for instance the transport of "renewable" Colorado River water pumped to Tucson and Phoenix, Arizona consumes huge amounts of resources, while its over-allocation has so depleted the river's flow, that vast tracks of the Colorado river delta and the culture of the indigenous people of the area have been destroyed.

Reservoir. A structure for storing water. It may be open or covered.

Retaining wall. A structure that holds back a slope, preventing erosion.

Rill. A tiny erosive drainage in which loose soil has washed away. It is very common on eroding slopes where roadways have been cut into hillsides or on bare dirt driveways and roads that run downslope.

Runoff. Water that flows off a surface when more rain falls than the surface can absorb.

Runoff coefficient. The average percentage of rainwater that runs off a type of surface. For example, a rooftop with a runoff coefficient of 0.95 indicates that 95% of the rain falling on that roof will runoff.

Runon. Runoff water that runs onto a site.

Saturated soil. Soil in which the pore space is completely filled with water.

Sediment. Soil, sand, and minerals washed from land into water or lower reaches of land, usually after rain. Excessive sediment can destroy fish-nesting areas; clog animal habitats, French drains, and porous pavement; and obscure waters so that sunlight does not reach aquatic plants.

Sewer. A pipe used to transport sewage elsewhere.

Sheet flow. The relatively even distribution of runoff water over the land surface, following the slope of the land downward, but not focused into distinct channels. Sheet flow has most likely occurred after a large rainfall if you don't see distinct channels in an area of sloping bare dirt.

Slope. A measurable steepness indicating a change in elevation from one point to another.

Soft water. Water containing little or no dissolved calcium and magnesium.

Solar access. Maintaining full winter sun exposure to winter sun-facing windows, solar water heaters, solar photovoltaic panels, solar ovens, and winter gardens.

Solar arc. A number of shading elements such as trees, cisterns, trellises, covered porches, and overhangs laid out in the shape of an arc or semi-circle open to the winter sun, and deflecting the rising and setting summer sun from any objects, such as a home or garden placed within the arc.

Spillway. A planned and stabilized route for overflow water.

Stormwater. Rainwater once it has landed on a surface.

Subsoil. The naturally compacted soil found beneath less compacted, more organic-matter-rich topsoil.

Subwatershed. A smaller watershed within, and making up part of a larger watershed.

Sun trap. An area having a more comfortable and moderate microclimate due the site being open to the winter's rising and noonday sun, while shaded from the afternoon sun—primarily in summer.

Superfund site. A site contaminated with hazardous substances, and approved for use of superfund trust money to fund cleanup efforts of the contamination.

Supplemental water. An auxiliary source of water meant to augment natural on-site rainfall resources.

Surface water. Water that flows on the surface of the land, such as water flowing in creeks and rivers.

Sustainable. A condition in which biodiversity and renewability of ecosystems, cultures, and natural resource production and quality are maintained over time.

Terrace. Sometimes called a bench, it is a relatively flat "shelf" of soil built parallel to the contour of a slope. The terrace reduces the steepness of a section of a slope, reducing runoff and erosion, while increasing infiltration. Terraces can be built with or without a retaining wall depending on the steepness of the slope.

Three-way diverter valve. A valve used to direct or divert water flow in one of two directions. Found at pool and spa supply stores, these valves can be incorporated into greywater plumbing to allow the user to send the greywater to the landscape or sewer as they please.

Tinaja. A desert water hole naturally carved into bedrock.

Topsoil. The upper layer of soil containing most of the organic matter and fertility.

Total dissolved solids (TDS). Indicates how many minerals and other solvents are contained in one gallon or liter of water. Technically, these are the dry residues that remain after the water has been heated to 180°C.

Toxic. Any substance able to cause injury to living organisms when eaten, absorbed through the skin, or inhaled into the lungs.

Transpiration. The loss of moisture from plants to the air via the stomata within their leaves.

Vent. A screened opening installed above the elevation of a closed water tank's inlet pipe to prevent a vacuum-caused implosion when large quantities of water are quickly drawn from the tank.

Wastewater. Water used by humans and considered a "waste" needing to be disposed of. Creating such a thing is the real waste.

Water softener. A device that replaces calcium and magnesium ions from hard water with sodium ions. Without the calcium and magnesium the water becomes "soft," but the added sodium or salt is not good for plants or soil. So softened water is not good for use with greywater systems, and the softener backwash is even worse since it has an even higher salt content.

Water table. The upper limit of a body of ground-water.

Watershed. The total area of a landscape draining or contributing water to a particular site or drainage.

Well. A human-made hole in the earth from which groundwater is withdrawn.

Wet system downspout. A wet system downspout drops to the ground where the horizontal run is supported by the soil in which it is buried, and then rises up again (but not higher than its inflow point at the gutter) to enter a tank or earthwork. The "wet" downspout is plumbed watertight because this section of pipe will continuously hold water. Sludge can collect within the watertight pipe, and the pipe is susceptible to freezing in winter if not adequately drained or installed.

Winter-sun side. For those living in the northern hemisphere, the south-facing side of buildings, walls, and trees is the "winter-sun side" and the north-facing side is the "winter-shade side." This is because the winter sun stays in the southern sky all day.

Index

More advance praise for

Rainwater Harvesting for Drylands
Volume 1

"Our modern society is afflicted with a severe case of hydrological illiteracy. Brad's Rainwater Harvesting for Drylands, in three volumes, is the antidote needed to mitigate this epidemic of cerebral imperviousness impacting the collective head-waters of our cultural ego-system! Pragmatic practitioners of "waterspread" restoration for arid lands and beyond, will find a wealth of accessible and practical information in these books. The Conservation Hydrology mantra of -Slow It - Spread It - Sink It – has never been better articulated in as clear and concise terms for the homeowner, ranch owner, sub-division developer or city stormwater engineer. I will require this book for all my *Basins of Relations* community watershed students and feel it should be so as well for all land managers and land use planners. This book is sure to become a classic for all people who believe in a future based on rehydration instead of dehydration and for that I say, Bravo Brad!"
— Brock Dolman, WATER Institute Director, Occidental Arts and Ecology Center

"Brad Lancaster presents the first of three volumes of his vision on what might be called Eco-hydrology. It is not just a book about harvesting rainfall, although there are many practical ideas on how to make use of the water that falls on your land. It is a guide designed to help the reader see what Brad sees when looking at a city lot or homesite. He has set as his goal to train you to see your land and the environment in which it is set in a new way, as a natural resource to be managed in harmony with your living there. While water harvesting is central to living a new paradigm it is only part of a broader vision designed to enrich your quality of life while enhancing the surrounding environment."
— James J. Riley, Ph.D., Soil, Water and the Environmental Science Department, The University of Arizona

"Like small acorns that grow into mighty oaks, Ben Franklin's succinct and wise words are perhaps more valuable today: 'Waste not; want not…A penny saved is a penny earned.' The anticipation of rain and its eventual harvest and storage for nurturing the native habitat, our source of food, the quality of our air and water, and visual delight for our senses is a natural model for us to mimic. Brad Lancaster's pioneering series of books, *Rainwater Harvesting for Drylands*, shows us how we can mimic the way nature works, as it immediately provides the resources necessary to support a world of efficient and effective use of water that helps create abundance in all our lives."
—Dr. Wayne Moody, American Institute of Certified Planners (AICP), Planner

More advance praise for

Rainwater Harvesting
for Drylands
Volume 1

"*Rainwater Harvesting for Drylands*, Volume 1 is more than a text on how to harvest rainwater. It is a way of life that gives back nourishment to our earth rather than what has become the standard of continually taking. This way of life has become almost a religion for Brad and as he demonstrates, it should be the same for all of us."

— Heather Kinkade-Levario, R.L.A., President of the American Rainwater Catchment Systems Association (ARCSA), Author of *Forgotten Rain - Rediscovering Rainwater Harvesting*

"The world needs more practical visionaries like Brad Lancaster! Blending his own knowledge, experience and wisdom with the collected wisdom and practices of water harvesters from around the world, Brad gives us access to a wealth of critically needed tools for rethinking our relationship with the gift of water from the sky. This man more than walks his talk; he lives, breathes, eats and drinks it!"

— David Eisenberg, Director of the Development Center for Appropriate Technology, co-author of the *Straw Bale House Book*, and a two-term member of the Board of Directors of the U.S. Green Building Council

"This book and the thinking behind it should be part of the basic education of civil engineers, architects, landscape architects and planners everywhere. As a civil engineer working for a progressive municipal water utility in an arid climate, I can see if a majority of our citizens followed these practices, many of our current and future challenges would be alleviated. The positive side benefits in terms of erosion-control, creation of bird habitat, and natural cooling would be exceptional."

—Patricia Eisenberg, P.E., Past president, Arizona Society of Civil Engineers

"On a water-world such as ours, Brad's book should be a required study for all human beings. Scholarly, forthright and, above all, practical, this work delivers critical knowledge to those thirsty for a positive relationship with water. Thank you Brad for your friendly presentation of such a complex and important component of global sustainability!"

—Paul A. Branson, Earthwise Technologies Ecological Restoration

The series continues...

Rainwater Harvesting for Drylands, Volume 2
Water-Harvesting Earthworks

Earthworks are one of the easiest, least expensive, and most effective ways of passively harvesting and conserving multiple sources of water in the soil. Associated vegetation then pumps the harvested water back out in the form of beauty, food, shelter, wildlife habitat, timber, and passive heating and cooling strategies, while controlling erosion, increasing soil fertility, reducing downstream flooding, and improving water and air quality.

Building on the information presented in Volume 1, this book shows you how to select, place, size, construct, and plant your chosen water-harvesting earthworks. It presents detailed how-to information and variations of all the earthworks, including chapters on mulch, vegetation, and greywater recycling so you can customize the techniques to the unique requirements of your site. Info on how to create cheap and simple tools to read slope and water flow are also included along with sample plant guides.

Real life stories and examples permeate the book, including:

- How homesteading grandmothers are restoring their land with simple earthworks made on their daily walks
- How curb cuts and infiltration basins redirect street runoff to passively irrigate flourishing shade trees planted along the street
- How check dams have helped create springs and perennial flows in once-dry creeks
- How infiltration basins are creating thriving rain-fed gardens
- How backyard greywater laundromats are turning "wastewater" into a resource growing food, beauty, and shade that builds community, and more.
- More than 225 illustrations and photographs.

For more details, publication dates, and ordering info see www.HarvestingRainwater.com.

Rainwater Harvesting for Drylands, Volume 3

Roof Catchment and Cistern Systems

Cisterns harvesting runoff from rooftop catchments have the potential to harvest the highest quality rainwater runoff on site, and allow for the greatest range of potential uses for that water. Water uses range from irrigation to fire protection, bathing, and washing, as well as potable water for drinking.

Building on the information presented in Volume 1, this book shows you how to select, size, design, build, or buy and install cistern systems harvesting roof runoff. Guiding principles unique to such systems are presented along with numerous tank options, and design strategies that will improve water quality, save you money, reduce maintenance, and expand the ways you can use your tanks and the water harvested within.

Additional information includes:

- How to size gutters and downspouts
- How to turn a drain pipe into a storage tank—how to build culvert cisterns
- How to retrofit ferrocement septic tanks for cistern use
- How to create a gravity-fed drip irrigation system for use with above-ground tanks
- How to create a pump-fed drip irrigation system for use with below-ground tanks
- System setup, maintenance tips, and water treatment options
- More than 60 illustrations and photographs

For more details, publication dates, and ordering info see www.HarvestingRainwater.com.